The World's Largest Producer of Organic Trace Minerals for Livestock

Alltech is the leading global animal health company providing all-natural nutritional solutions that benefit animal health and performance, setting the standard in animal nutrition worldwide since 1980.

Healthier animals are better able to resist disease challenges, handle production stress, have higher reproductive performance and gain weight more efficiently. With the proven success of brands such as Yea-Sacc®[1026], Sel-Plex®, Bio-Mos®, *Mycosorb®, MTB-100®, Bioplex® and Allzyme®Vegpro™, the company's global product line is a unique example of how all-natural technologies backed by continuous research can help move the industry forward.

Research

Alltech is recognised as one of the leading research based animal health companies in the world. With our commitment to research and development, our successful programme of marketing through education and our pioneering ACE concept, we are ensuring all our products are safe for the **A**nimal, the **C**onsumer and the **E**nvironment. Every new Alltech solution is measured against this programme. Every new Alltech solution has its origins in nature… *naturally.*

*not available in the United States

Alltech Bioplex® Production Facilities Assure Premium Quality

The unique process that creates Bioplex® organic trace minerals takes place in equally unique production facilities. Each Alltech facility uses enzymes produced by Alltech and non-animal protein sources to produce chelated trace minerals that are stable and consistent from batch to batch.

How do we do it?

The manufacture of Bioplex® minerals is carefully controlled. Routine quality assurance tests measure the percent of mineral and the degree of chelation using state-of-the-art techniques. Only when Alltech's strict standards are met will the finished product be released from quarantine and shipped to over 85 countries worldwide.

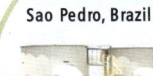

Alexandria, Canada

Housing three spray driers, this facility has the capacity to produce 20,000 tonnes of Bioplex® minerals each year.

Sao Pedro, Brazil

Home to the largest spray drier in the world capable of producing 60 tonnes per day.

BIOPLEX®
Redefining Mineral Nutrition

Form Follows Function: *following nature's food chain*

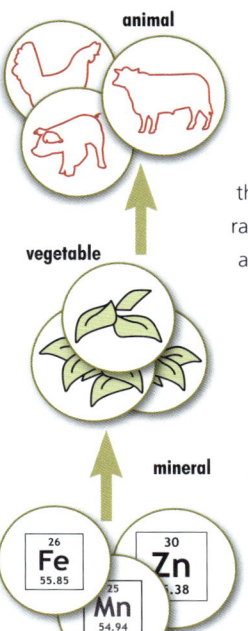

Alltech's Bioplex® organic trace minerals provide trace mineral nutrition in a form as close to nature as possible. By presenting minerals in the same types of compounds found in grains and forages, Bioplexes® are better able to meet the higher nutrient needs of modern livestock for rapid growth, maximum reproductive efficiency and animal health.

Immunity

Easily absorbed from the intestine, Bioplex® minerals are more highly retained in the animal. Improved mineral status has been shown to have a positive effect on immune function in all livestock species.

Bioplex® minerals and nutrient management

There is increased concern about environmental contamination due to excess dietary minerals being excreted by production animals. Due to their higher bioavailability, Bioplex® minerals are better retained and improve animal mineral status while reducing excretion levels, thus being beneficial to both the animal and the environment.

BIOPLEX®

Redefining Mineral Nutrition

Providing Trace Mineral Nutrition in a Form as Close to Nature as Possible

Alltech's unique Bioplex® organic trace mineral production process begins in our own facilities using natural enzymes. All raw materials and finished products are tested before and after production to ensure the absence of contaminants such as dioxin and heavy metals. This is done using the most stringent global quality guidelines.

Bioplex® minerals are unique in that they are protected by multiple amino acids and peptides rather than single amino acid-linked compounds, replicating the form found in plants. This level of protection gets them safely through the volatile GI tract, making them more easily absorbed and more highly retained.

Each batch of Bioplex® minerals is quarantined prior to undergoing Alltech's unique Quality Assurance analysis to assess chelation, ensuring every product has the percent total and bound mineral content required. Once the product has met Alltech's strict standards it is packaged and shipped to 85 countries worldwide.

Bioplex® minerals are available as individual minerals or in blended packs. The range includes*:

- Bioplex® Cobalt
- Bioplex® Copper
- Bioplex® Iron
- Bioplex® Magnesium
- Bioplex® Manganese
- Bioplex® Zinc
- Bioplex® High 4 - Formula
- Bioplex® Quadra - Formula
- Bioplex® Bovine - Formula
- Bioplex® Pig - Formula
- Bioplex® Poultry - Formula
- Bioplex® ZMC 842

*Not all products are available in every country. Contact your local Alltech representative for details.

SEL-PLEX®

Nature's Form of Selenium

Sel-Plex® is Alltech's proprietary selenium-enriched yeast. It enhances health and performance by providing a more bioavailable form of selenium that the animal can absorb, retain and utilise for metabolic processes far more efficiently than inorganic selenium.

As we demand more from our animals, we must demand more from our minerals

Selenium is involved in a number of important systems within the body, including antioxidant defence, reproduction and immune function. Because Sel-Plex® is better retained, the animal builds up nutrient reserves for periods of increased antioxidant demand and challenge without risking toxicity.

Proven safety and performance

Sel-Plex® is the most proven form of selenium available with the only EU approval and FDA allowance.

Sel-Plex® has been reviewed and accepted for use in the following diets: broilers, breeders, layers, turkeys, pigs, dairy cows, beef cows, horses, sheep, and dogs.

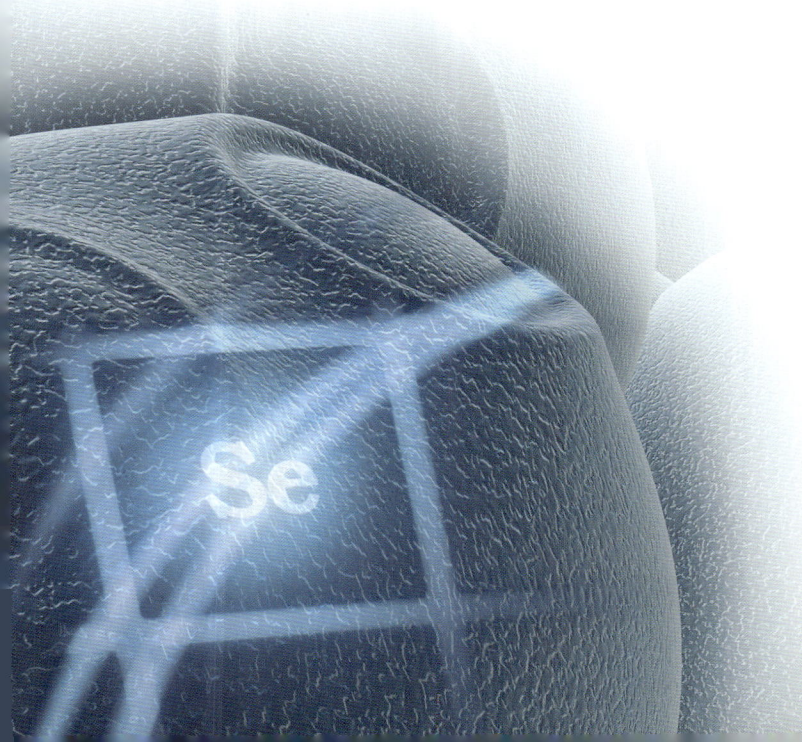

SEL-PLEX®

Sel-Plex® is 3x more Bioavailable than Sodium Selenite

- Even at maximum allowable levels in production diets, sodium selenite typically does not meet the animal's needs.
- Selenium is a nutrient intended for use in the body's antioxidant defences. Sodium selenite is selenium in its most highly oxidised and toxic form, leading to oxidative consequences.
- Sodium selenite cannot be converted to the amino acid selenomethionine, which is the way in which selenium is stored in the animal's tissues.

Nature's way is the best way

Sel-Plex® is synthesised by growing a specific strain of yeast on media that is deficient in sulphur and has a high concentration of selenium. Because sulphur and selenium are chemically similar, selenium-supplemented yeast synthesises selenomethionine and selenocysteine instead of forming the sulphur containing amino acids methionine and cysteine. The synthesis of these selenoproteins is important because Sel-Plex® uses the same storage method found in the plants that livestock evolved eating, and, therefore, it is the best way to supplement selenium in domesticated species.

Sel-Plex® * is available in a range of concentrations including, Sel-Plex® 600 ppm, 1,000 ppm and 2,000 ppm.

*Not all concentrations are available in every country. Contact your local Alltech representative for details.

Profile

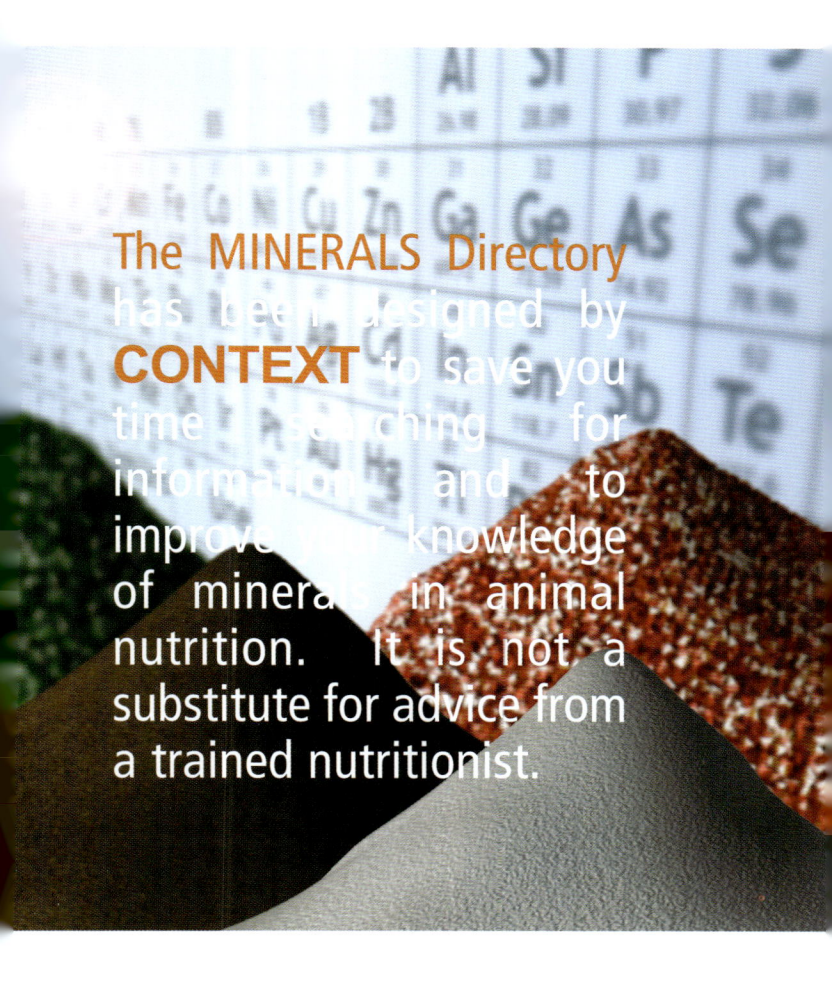

The MINERALS Directory has been designed by **CONTEXT** to save you time in searching for information, and to improve your knowledge of minerals in animal nutrition. It is not a substitute for advice from a trained nutritionist.

About the Authors

Dr Wesley Ewing BSc (Hons) Agric, Dip M, MSc, PhD.

After the international success of The FEEDS Directory, by Dr Ewing, it proved the need for further easy to use guides on animal nutrition and especially minerals. Wesley has both a commercial and technical knowledge of animal nutrition having worked in the industry for 20 years for global leaders such as Cargill and Provimi. He was the General Manager for a UK animal nutrition and health company before establishing Context.

Sally J. Charlton BSc (Hons) Ani Sci, Pr Sci Nat.

Mrs Charlton has broad knowledge of mineral nutrition having worked in managerial and advisory positions in the animal feed industry for the last 15 years. She was a product manager for Trouw Nutrition, a leading European manufacturer of premixes and speciality feed. She was a member of the South African Council for Natural Scientific Professions and has been a consultant animal nutritionist in South Africa and Canada over the last 7 years.

CONTEXT   www.contextbookshop.com

The Minerals Directory

Title: The MINERALS Directory: 2nd Edition
Your easy to use guide on minerals in animal nutrition

1st Published 2005, 2nd Edition Published 2007

British Library Cataloguing in Publication Data
The MINERALS Directory (2nd Edition)
I.Ewing, W.N.   2.Charlton, S

ISBN 978-1-899043-11-8

© Context 2007

All rights reserved. No part of this publication may be reproduced, in any material form (including photocopying or storing in any medium by electronic means and whether or not transiently or incidentally to some other use of the copyright holder) except in accordance with the provisions of the Copyright, Designs and Patents Act 1998. Applications for the copyright holder's written permission to reproduce any part of this publication should be addressed to the publishers. Whilst every effort has been made to ensure the contents are correct, the author and publisher cannot be held responsible for any errors or omissions contained herein.

Produced and published by Context
Context Products Ltd
53 Mill Street
Packington
Ashby de la Zouch
Leicestershire
England
LE65 1WN
Tel: 01530 411337
Fax: 01530 411289
Email: enquiries@contextproducts.co.uk
www.contextproducts.co.uk

Disclaimer:
No responsibility is assumed by the publisher for any injury and/or damage to persons or property as a matter of products liability, negligence or otherwise, or from any use of any instructions, data or ideas contained in the material herein. Because of rapid advances in genetic development, country variation and analytical differences data will vary from that listed. The authors and CONTEXT accept no liability for errors in this guide. We recommend that all data is checked from at least two other independent sources.

Information on minerals is widespread and it has not been possible to review all references. This guide is not intended to replace the many general or more specific nutritional texts currently available. Continual revision and updating is planned as new information becomes available. We would welcome your support on updating this guide. Email wewing@contextproducts.co.uk

Finally this is a colour guide. If you receive a Black and White version please inform us of the copyright theft at the address above.

Animal Nutrition and Farming Publications

www.contextbookshop.com

CONTEXT   www.contextbookshop.com

The Minerals Directory

Content

	Section
Introduction	A
Abbreviations	B
Conversions	C
Interpretation of sections	D
Methods of feeding minerals	E
Organic minerals	F
Periodic table	G
Water quality	H
Aluminium	1
Arsenic	2
Boron	3
Cadmium	4
Calcium	5
Chloride	6
Chromium	7
Cobalt	8
Copper	9
Fluorine	10
Iodine	11
Iron	12
Lead	13
Magnesium	14
Manganese	15
Mercury	16
Molybdenum	17
Nickel	18
Phosphorus	19
Potassium	20
Selenium	21
Silicon	22
Sodium	23
Sulphur	24
Tin	25
Vanadium	26
Zinc	27

This is a limited edition version of The Minerals Directory personalised for **Alltech**

CONTEXT   www.contextbookshop.com

Introduction

General

Mineral nutrition is a small part (in number) of the complete nutrition of an animal. It can often be ignored and yet it is critical to the well-being and top performance of every animal after energy and protein needs have been met. The need to assess the mineral balance of animals has become more important for a number of reasons:

- Higher production levels from high genetic merit livestock
- Changes in feedstuffs – global availability, processing methods etc
- Reduction in use of animal co-products (often good sources of minerals)
- Changing soil balances, crop fertilisation and varieties etc
- Complex interactions of mineral elements
- Changes in husbandry practices – bedding materials, housing etc
- The need to reduce environmental damage from over-use of chemicals
- Newer supplements affecting methods of absorption and availability

The aim of this Directory is to provide a functional, easy to read review of minerals in animal nutrition. Requirements have been summarised in relation to NRC publications. Commercial allowances will vary between countries and within countries due to the inherent status of minerals in that area, feeds utilised and genetic and production requirements of the animals. Some figures have been included but local nutritionists can provide additional information and in relation to specific area and farm requirements.

Mineral nutrition is dynamic and requires consistent, regular and accurate assessment in all animal production systems.

Minerals

Minerals are inorganic elements required in small quantities for the normal growth and reproduction of animals. All animal tissues contain inorganic or mineral elements. They are frequently found as salts with either inorganic or organic compounds. Minerals are present in all feedstuffs and are the inorganic components that remain after burning, e.g. ash.

Analysis

Ash is used as an indicator of total mineral level of feeds. A sample is burnt at a temperature of between 350 and 600°C to destroy all organic material. Total ash values are useful in the proximate analysis of feeds and to estimate the dust and soil contamination in harvested forages. It is not a good indicator of the useful mineral content of feeds or for expressing mineral requirements. Ash level of feeds is not 100% accurate as:

- Volatile material e.g. chlorine, iodine and selenium are lost on burning
- It does not indicate which of the minerals it contains or relative amounts
- Actual weight includes other components which are in the mineral constituent e.g. the oxide, carbonate, sulphate component

Atomic absorption spectrophotometry (AAS), and neutron activation analysis (NAA) are the most frequently used analytical techniques. Other techniques include, proton induced x-ray emission (PIXE), isotopic dilution mass spectrometry (IDMS), x-ray fluorescence (XRF), inductively coupled plasma optical emission spectroscopy (ICP-OES), and near infrared reflectance spectroscopy (NIRS). Sampling, storage, handling and analytical techniques must ensure minimal contamination.

Analytical values of individual mineral elements do not take into account mineral digestibility that may affect relative bio-availability for different species and sources.

Introduction

Classification of Minerals

Classification of Minerals
There are over 60 elements found in soils, which are taken up by plants. For animals, there are 27 essential minerals. Essential minerals are required for maintenance and to support adequate growth, reproduction and health. Non-essential minerals are widely found in the earth but are not thought to play an active role in the body.

Essential Macro Minerals (Major) (7)
Found in the diet greater than 100ppm. (Levels usually % or g/kg)

Calcium	Ca
Chlorine	Cl
Magnesium	Mg
Phosphorous	P
Potassium	K
Sodium	Na
Sulphur	S

Essential Micro Minerals (Trace) (9)
Found in the diet at less than 100ppm. (Levels usually ppm or mg/kg)

Chromium	Cr
Cobalt	Co
Copper	Cu
Iodine	I
Iron	Fe
Manganese	Mn
Molybdenum	Mo
Selenium	Se
Zinc	Zn

Essential Minor Minerals (11)
(Beneficial in some circumstances)
Some known for their toxic effects. Levels usually as mg/kg or µg/kg(ppb)

Aluminium	Al
Arsenic	As
Boron	B
Bromine	Br
Fluorine	F
Lithium	Li
Nickel	Ni
Rubidium	Rb
Silicon	Si
Tin	Sn
Vanadium	V

Highly Toxic Minerals (3)
(Non essential)
Levels usually mg/kg or µg/kg(ppb)

Cadmium*	Cd
Lead*	Pb
Mercury*	Hg

In addition to these minerals, plant and animal tissues can also contain 20-30 other minerals but usually in small and variable quantities. No function has so far been found for these accidental minerals but they are most likely from the environment and diet in which the animal lives. For example, Lithium, Beryllium, Scandium, Gallium, Germanium, Zirconium, Silver, Antimony, Caesium, Barium, Bismuth, Radium, Thorium, Uranium.

Introduction

Minerals

Mineral Function
Single elements are not often independent or self sufficient in their role in body processes. Functions are most often interrelated and balanced against each other. Typically functions for macrominerals and micro minerals are listed below;

Macrominerals
- Structural components of the body e.g. bone and tissue
  (e.g. muscle, organs, blood, soft tissue)
- Body fluid components
- Key role in maintenance of osmotic pressure and acid-base balance
- Critical for nervous transmission and membrane electric potential and muscle function

Microminerals
- Components of hormones in the endocrine system
- Components of metallo-enzymes and enzyme factors
- Components of some vitamins and amino acids
- Regulatory, control of cell replication and differentiation
- Related to adequate immune response

Sources
Minerals enter the body naturally from ingestion of food and water. An ingestion shortage leads to a shortage in the body that can lead to deficiency while an excess can lead to toxicity. There is a range in between under which the body can operate. The level of minerals in feedstuffs varies greatly and some reliable levels are hard to find. The feedstuff may also vary, even within a batch, due to different growing conditions, soil type, plant species and fertiliser application, environment, and maturity. The levels in co-products will also be affected by the processing method.

Inorganic compounds, often used to supplement feed levels, come from the earth or industrial manufacturer. For trace minerals, their chemical and physical forms are probably the most significant factors affecting digestibility and availability.

Some minerals are present in the diet in quantities that meet the production demands of that species and so do not require supplementation. Each mineral is assessed for its adequacy to meet requirements. The level of iron, copper, zinc, cobalt, manganese, iodine and selenium found in feed raw materials is often too low to meet production requirements and so must be supplemented.

Availability
Chemical analysis of feedstuffs does provide levels for mineral elements, however, not all this is available to the animal. During digestion the form in which the mineral is bound up will determine the utilisation. Following digestion the nutrients are absorbed into the blood with a further loss. How much can be absorbed from the gut depends on species and age of animal, intake vs requirement, chemical form, synergists and antagonists in the diet and environmental factors. Interaction between minerals can affect digestion and absorption. The animals own regulatory system can also mean that only the body requirements' are absorbed.

Introduction

Minerals

Absorption and Metabolism

Digestion and absorption of mineral elements by an animal provides an estimate of its bioavailability.
Apparent absorption = (intake − total faecal excretion/ intake) x 100
This is limited to elements where faeces is the major pathway of excretion (e.g. Ca, P, Zn, Mn, Cu)
True absorption is same as apparent but also takes account of endogenous faecal excretion losses. This value is a more valid estimate of the amount available for physiological purposes.
Minerals are converted to their free form (ionic state) in the gut and absorbed across the gut wall by either active or passive means.

Active Absorption

This is when the element must be forced by the intestinal wall from the lumen into the cells of the body e.g. calcium, phosphorous and sodium. Often this is against a concentration gradient using energy.

Passive Absorption

This is the usual method for most minerals to cross the gut wall going from a higher to lower concentration.

Factors Affecting Absorption

Antagonists such as phytates, oxalates and fat can bind certain minerals and make them less available. Minerals can also interfere with the utilisation of other minerals (see mineral interrelationships). Absorption tends to be more efficient in younger animals and is affected by gut pH and the form ie organic vs inorganic. Usage of ingested and absorbed minerals is not 100% as there is wastage at the kidneys, digestive system and skin that must be replaced.

Requirements/ Nutrient Demand

Each animal has different dietary requirements that are directly linked to animals' physiological state, size, sex, age, species, breed, genetics, diet, chemical form of ingested elements, productivity etc.
Demand usually decreases, as an animal grows older although absorptive ability will also reduce, thus diet levels may not decrease in proportion.

Maintenance

This is the level of mineral needed to keep the existing body tissues maintained and aid in the body functions such as digestion, circulation and respiration. As production level and food intake increases maintenance level will rise, as the need to metabolise these nutrients will increase body function. Maintenance requirements often include endogenous, urinary and sweat loses.

Production

Minerals are required for every gram of production e.g. liveweight gain, milk yield, egg production, wool production etc.
This requirement can vary depending on the animals' growth cycle, production level and quality. For example, young animals need to lay down more bones while older animals are generally depositing a higher proportion of muscle. Bones need elements such as calcium and phosphorous for structural components.

Reproduction

This is the amount required for foetal growth and maintaining the placenta, uterus and uterine contents. Greatest levels are required during late gestation when in some circumstances food intake and therefore supply may be temporarily reduced. A shortfall can cause sterility, low fertility, silent oestrus or failure to get pregnant.

Introduction

Minerals

Growth
Different organs grow at different times and have different mineral requirements. Increase in body weight is usually early in life.

Immune Function
Relatively new information describes the effect of nutrition on immune function. It is thought optimal immune responsiveness and disease resistance is greater than needed for normal growth. Se, Cu, Zn, and Co deficiencies have been shown to alter various components of the immune system.

Priority of Demand
Minimum requirements for one function may be higher than another. This can affect mineral levels required for adequate status of production.
E.g. Mn requirement for growth are lower than for fertility.

Adequate Intake
Requirements for maintenance, production, reproduction and growth must be satisfied by dietary intake. The diet should provide sufficient minerals allowing for absorption losses, as not all will be available or fully utilised. The diet is defined as feed and water intake.
Mineral requirements in this publication have been expressed as a proportion of the dry matter consumed. They can also be expressed in other ways, e.g. amounts per day, amount per unit of product (e.g. milk, eggs). A range of levels, often indicates differences in liveweight or production demands. Adequate levels are difficult to establish and are mainly based on the minimum level to avoid a deficiency. Tables are not always based on modern genotypes. A daily allowance is often used as the requirement to avoid deficiency and a level to optimise performance. Mineral intake may need to be sufficient to ensure long term maintenance of body mineral reserves and levels in edible products of the animal. The animals body can reduce the level of minerals in its products to try and maintain body levels under suboptimal intakes.

Deficiency and the Signs
Deficiency levels depend on the animal, intake level, duration, antagonisms, age, sex, species etc. Many animals suffer from deficiencies of nutrients that are never seen. Deficiencies range from marginal (sub clinical) conditions to severe disorders. e.g.. metabolic conditions.
Marginal deficiencies of minerals often reflect in reduced food intake, digestion, absorption, performance, and may even be mistaken in some countries for parasite infestation. Livestock can usually cope with a small deficiency for a short period easily, by reducing the mineral levels in the wool, eggs and other products except for milk that usually maintains the same concentration. Marginal deficiencies may appear small but can cause substantial economic losses. Marginal deficiencies under low levels of production, become more important and noticeable as production level increases.

Toxicity/ Tolerance
While minerals may be essential for growth, excess can be damaging to the animal. Animals can be inadvertently fed an excess of a mineral without a toxicity being seen.
Toxicity depends on the mineral, level, duration and antagonists. Tolerance levels obtained for one species are often extrapolated to other species.

Introduction

A6

Minerals

Supplementation
The simplest form is usually through the diet. The mineral supplement must be palatable and of low volume to avoid a reduction in the intake of other nutrients. Mineral supplementation is usually from either inorganic salts or combined with organic mineral complexes.
Inorganic supplements are derived from non living sources usually dug out of the ground e.g. limestone. Organic minerals are bound with animal or vegetable material. Metal chelates such as copper, zinc, manganese are usually found as metal amino acid complex, metal amino acid chelate, metal proteinate, or metal polysaccharide. These are thought to increase the availability of specific trace elements.

The increase in organic farming is increasing the focus on natural sources of supplementation. E.g. yeasts, algae, plant or herb extracts. Consistent analysis is required but there may also be an economic penalty.

Mineral Interrelationships
A mineral element can interact with and influence the requirement for itself and other elements. The relationship of an excess or an inadequate amount of a specific mineral or other nutrient upon the utilisation or availability of others in the diet, is one of the most intriguing phases of mineral nutrition. The figure below diagrammatically demonstrates the complexity of the situation. This is only a partial representation as many of the newer minor and toxic minerals are not included. This is a topic for discussion in itself and until mineral interrelationships are fully understood, only an approximation of mineral requirements can be made.

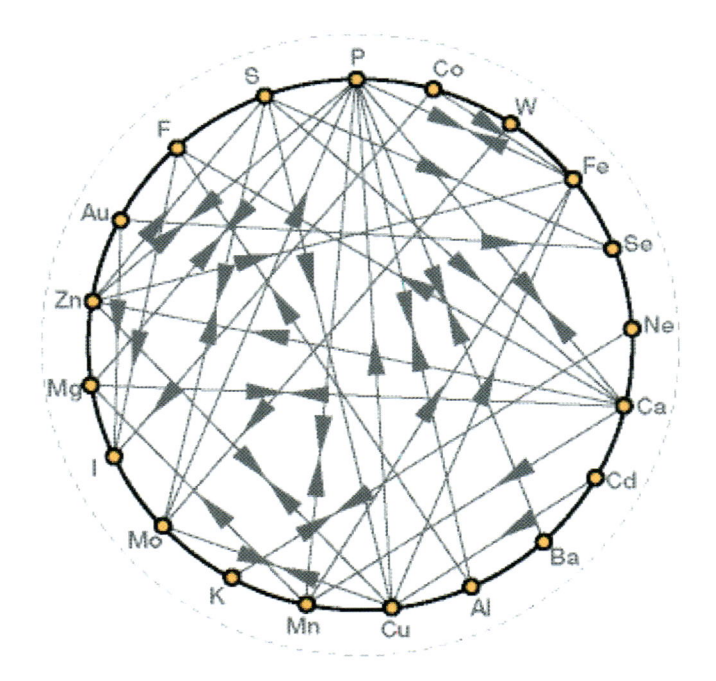

CONTEXT www.contextbookshop.com

Abbreviations

B1

Useful Terms and Abbreviations

DNA	Deoxyribose nucleic acid
DM	Dry Matter
g%	Grams per 100ml
g/tonne	Grams per tonne (=ppm)
GIT	Gastrointestinal tract
GSH_Px	Glutathione peroxidase
Hb	Haemoglobin
IU	International unit
l	Litre
mcg	Microgram
mEq	Milli equivalents
mg%	Milligrams per 100ml
mg/kg	Milligrams per kilogram (=ppm)
mmol	Millimoles
MR	Milk replacer
NaCl	Sodium chloride (salt)
NDF	Neutral detergent fibre
ng/g	Nanograms per gram (=ppb)
nmol	Nanamoles
pmol	Pico moles
ppb	Parts per billion
ppm	Parts per million
ppt	Parts per trillion
PTH	parathyroid hormone
RNA	Ribonucleic acid
SOD	Superoxidase dismutase
T_3	Triiodothyronine
T_4	Thyroxine
μ%	Micrograms per 100ml
μg/kg	Micrograms per kilogram (=ppb)
μg/g	Micrograms per gram (=ppm)
μmol	Micro moles
WW	Wet weight
<	Less than
>	Greater than

CONTEXT

www.contextbookshop.com

Conversions

C1

Molar Conversion Factors

Element	Symbol	From	Multiply by	to
Aluminium	Al	µg/l	37.06	nmol/litre
Arsenic	As	ppm	13.35	µmol/litre
Boron	B	ppm	92.50	µgmol/litre
Cadmium	Cd	µg/100ml	0.08897	µmol/litre
Calcium	Ca	mg/100ml	0.2495	mmol/litre
Chlorine	Cl	mEq/l	1	mmol/litre
Chlorine	Cl	mg/100ml	0.2821	mmol/litre
Chromium	Cr	ppm	19.23	µmol/litre
Cobalt	Co	ppm	16.9684	µmol/litre
Copper	Cu	ppm	15.74	µmol/litre
Copper	Cu	µg/100ml	0.1574	µmol/litre
Fluorine	F	ppm	52.63	µmol/litre
Iodine	I	µg/100ml	78.80	nmol/litre
Iron	Fe	µg/100ml	0.1791	µmol/litre
Lead	Pb	µg/100ml	0.04826	µmol/litre
Magnesium	Mg	mg/100ml	0.4114	mmol/litre
Magnesium	Mg	mEq/litre	0.5	mmol/litre
Manganese	Mn	ppm	18.20	µmol/litre
Mercury	Hg	µg/100ml	49.85	nmol/litre
Molybdenum	Mo	ppm	10.42	µmol/litre
Nickel	Ni	µg/litre	0.01703	µmol/litre
Phosphorus	P	mg/100ml	0.3229	mmol/litre
Potassium	K	mEq/litre	1	mmol/litre
Potassium	K	mg/100ml	0.2558	mmol/litre
Selenium	Se	ppm	12.665	µmol/litre
Silicon	Si	ppm	35.60	µmol/litre
Sodium	Na	mEq/litre	1	mmol/litre
Sodium	Na	mg/100ml	0.4350	mmol/litre
Tin	Sn	ppm	8.425	µmol/litre
Tungsten	W	ppm	5.439	µmol/litre
Vanadium	V	ppm	19.63	µmol/litre
Zinc	Zn	ppm	15.30	µmol/litre
Zinc	Zn	µg/100ml	0.1530	µmol/litre
Ammonia	NH_3	µg/100ml	0.5871	µmol/litre
Carbon dioxide	CO_2	mEq/l	1.0	mmol/litre
Glucose	Glu	mg/100ml	0.05551	mmol/litre
Haemoglobin	Hb	g/100ml	0.6206	mmol/litre

ppm (µg/g or mg/kg) x mole (mol.wt.) = mmol/litre or kg

ppm x (1000/mole) = µmol/litre or kg

CONTEXT www.contextbookshop.com

Conversions

Metric Weights and Measures to Imperial

Units From	Conversion Factor	Units To
Length		
mm	x 0.04	ins
cm	x 0.4	ins
mm	x 1.1	yds
km	x 0.62	miles
Mass and Weight		
g	x 0.03527	ozs
g	x 0.002205	lbs
kg	x 2.2046	lbs
g	x 0.00422	cups
kg	x 4.2	cups
metric ton	x 1.102	short tons
metric ton	x 0.984	long ton
metric ton	x 2204.6	lbs
mg/kg	x 1	ppm
iu/kg	x 0.454	iu/lb
Volume		
ml	x 0.0338	fl.oz
litre	x 33.81	fl.oz
litre	x 2.1134	pints
litre dry	x 0.908	quart dry
litre	x 1.057	quart liquid
litre	x 0.2642	gallons
litre	x 4.166	cup
Capacity		
cm^3	x 0.061	cubic in
m^3	x 35.315	cubic ft
m^3	x 1.308	cubic yd
Temperature		
°C	x (9/5)+32	°F
°F	-32 x (5/9)	°C
Energy		
kcal/kg	x 0.454	kcal/lb
mcal/kg	x 0.454	mcal/lb
MJ/kg	x 0.24	mcal/kg

CONTEXT   www.contextbookshop.com

Conversions

C3

Common Conversion Factors (Systeme international d'unites) (SI Units)

From	to	Multiply by
g/l	oz/gal	0.1335
	mg/ml	1
	mg/l	1000
	wt%	0.1
mg%	μg%	1000
	mEq/l	10/(atomic mass/valency)
	mg/l	10
mEq/l	mg%	0.1 x (atomic mass/valency)
	μg%	100 x (atomic mass/valency)
	mg/l	(atomic mass/valency)
mg/l	mg%	0.1
	μg%	100
	mEq/l	1/(atomic mass/valency)
	g/l	0.001
	oz/gal	0.0001335
	molar	1/1000 x atomic mass
mg%	mg%	0.001
	mEq/l	0.01/(atomic mass/valency)
	mg/l	0.01
molar	mg/l	1000 x atomic mass
oz/gal	g/l	7.491
	mg/l	7491

CONTEXT   www.contextbookshop.com

Interpretations

D1

Essential Notes for this Guide

Additional interpretation for sections within each mineral element:

Requirements

Based on NRC references: 100% DM diets unless marked * then 90% DM diets)
Dairy – 2001, Beef -1996, Sheep- 1985, Horse- 1989, Swine- 1998*,
Poultry-1994*, Fish-1993*, Dogs-1985a, Cats-1986, Rabbits- 1977*
NRC requirements are often based on growth performance and quantities of a specific mineral sufficient to prevent clinical signs of a deficiency.
Note Figures quoted for Cats are for growing kittens.

Allowances

Examples of some typical allowances have been included, and will vary within species and for different growing areas.

Adequate Status

Confirmation of guidelines should be obtained from reporting laboratory due to varying methodology. Care must be taken to compare same units.
Most tissue levels are on a wet weight basis in ppm.
Wet weight x 3.5-4.0 = approximate dry weight for most species
Wet weight x 5.0-6.0 = approximate dry weight for foetal tissues

Kidney levels refer to kidney cortex.
More than one mineral may be involved, interactions are common and should not be overlooked.
Please note maximum dietary tolerable levels have been taken from NRC 1980.

Deficient

The level at which clinical or pathological signs of deficiency should be apparent.
<u>Acute</u>- condition appears rapidly and follows a short, severe course.
<u>Chronic</u>- condition lasts for a long period of time or marked by frequent recurrence

Marginal

The level at which sub clinical effects may occur, for example reduced immune response or growth rate.

Adequate

The level sufficient for optimum functioning cf all body mechanisms with a small margin of reserve to counteract commonly encountered antagonistic conditions

Toxic

The level at which subclinical, clinical or pathological signs of toxicity would be expected to occur

Normal

Sometimes used where deficiency is unknown and indicates normal background level

Antagonism

An excess of one mineral or minerals can interfere with the uptake or metabolism of another mineral

Synergy

Mineral elements that work together or substitute for one another in their function

CONTEXT   www.contextbookshop.com

Interpretations

Essential Notes for this Guide

Maximum Tolerable Level

The "maximum tolerable level" is defined as "..that dietary level that, when fed for a limited period, will not impair animal performance and should not produce unsafe residues in human food derived from the animal." Where possible, these levels were generated for cattle, sheep, swine, poultry, horses, and rabbits on the basis of the best information available. Where it was considered appropriate, the data generated with one species was used in estimating a tolerance level for another. Good nutritional practices dictate that, in general, dietary levels of the mineral elements should be well below the maximum tolerable levels. With elements such as lead, cadmium and mercury, dietary levels should be maintained as low as possible in order to minimise the carryover of these elements into the human diet.

Biological Availability

Biological availability or utilisation of trace mineral elements is influenced or affected by several factors and very complex. A partial listing would include (not necessarily in order of importance): Chemical form of the element, animal species, age, sex, nutritional status, levels and forms of other elements, level and type of other nutrients (e.g. protein), animal health disease, parasite infestation, genetics, homeostatic control, hormone control, environmental factors, chelating agents, feed processing, type of diet or ration fed.

Effective methods for determining availability vary greatly among; different minerals and animal species. Conventional determinations of differences between intake and fecal excretion do not establish availability for many minerals because it is impossible to distinguish between unabsorbed and endogenous minerals. Radioactive isotopes can be effective for some elements but even with isotopes there are problems. Some availability data can be unreliable due to ineffective experimental procedures.

Feed Analyses

Nutrient content of a range of feed ingredients based on averages from ongoing databases. Variation in analyses will occur due to growing conditions, type of processing, sampling etc. The most reliable is analysis of local area ingredients and forages or routine samples of feed materials being fed.

These are analytical values and do not take into account mineral digestibility or interactions that may affect relative bioavailability to different species. Users of this guide should note that where no level is shown there has been insufficient data found by the authors. We will update the guide as information is sourced.

Methods of Feeding

Methods of Feeding Minerals to Livestock

Most minerals are not very palatable to animals, except salt. Salt is often used as a carrier for other minerals and to encourage intake. Other appetite stimulators include

Diet Supplementation
Mix specific mineral premix or supplement into the feed at required concentration

Top Dressing
Mineral supplement is 'top dressed' over feeding area such as silage, feeding troughs etc.
Top dressing can allow selection by stock and uneven intakes

Free Choice Supplementation
- Free access minerals offered as a single complete mixture
- Salt licks (compressed and containing 60 to 98 per cent sodium chloride). The trace element content is variable and some of the major elements e.g. phosphorus may be included in those with lower salt content
- Molasses or syrup licks. Intake is restricted by rotating wheel or balls sitting on the surface of the liquid
- Compressed blocks. Containing cereals, molasses and other feeds as well as urea, an energy source and minerals. Intakes need to be controlled by using containers
- Mineralised nuts are frequently treated with silicones to improve their resistance to crumbling in bad weather.

Consumption can vary between animals in a group and some minerals can be lost from volatilisation or leaching.
Level of mineral in mixture, daily intake and availability of source to meet daily recommendation is required.

Minerals in Water
Knowledge of water quality and quantity is required. Intakes can vary between animals and from environmental conditions. Only water soluble minerals can be supplemented.

Drenches or Injections
Can be used to provide some minerals but not satisfactory for those required on a daily basis or where stored element is not readily available.
Injectibles can help prevent and cure some deficiencies. E.g. Iron for piglets.

Ruminal Preparations
Sustained, slow release formulas for one or more specific minerals in the rumen. Can be regurgitated and coated with impervious layer (e.g. calcium phosphate) in rumen. Release is usually by diffusion, dissolution or by galvanic effect.

Organic Minerals

F1

Chelates

There are a number of organic trace mineral products on the market and the information available on their action has increased dramatically over the last 10-15 years. This summary defines and provides basic information on chelates and mode of action but not specific products.

In simple terms, a chelate is a mineral or metal atom chemically bonded to an organic molecule (chelating agent). This chelate can be amino acids, di- and tripeptides

The word 'Chelate' comes from the Greek word *'chele'* meaning claw, which is appropriate for the way in which the cations are held by the metal binding agents. The organic substances which bind to the metal are called ligands.

Organic chelates of minerals are important factors influencing absorption of these elements. The chelate carries no electrical charge, and as such is stable through the pH changes that take place during digestion. Being insoluble, the mineral is in a form to be absorbed into the bloodstream. The ideal chelating agent is one that releases the mineral in the ionic form at the intestinal wall, or that can be absorbed as an intact chelate.

AAFCO definitions for organic mineral complexes are:

Metal Amino Acid Complex
Product resulting from the complexing of a soluble metal salt with an amino acid. Minimum metal content must be declared. Metal (specific amino acid) Complex: same as above but amino acid is specified.

Metal Amino Acid Chelate
Product resulting from the reaction of a metal ion from a soluble metal salt with amino acids with a ratio of one mole of metal with one to three (preferably two) moles of amino acid to form coordinate covalent bonds. The average weight of the hydrolysed amino acids must be approximately 150 and the resulting molecular weight of the chelate must not exceed 800. Minimum metal content must be declared.

Metal Proteinate
Product resulting from the chelation of a soluble salt with amino acids and/or partially hydrolysed protein. Must be declared as an ingredient as the specific metal proteinate.

Metal Polysaccharide Complex
Product resulting from the complexing of a soluble salt with polysaccharide solution. Must be declared as an ingredient as the specific metal complex.

AAFCO does not have an approved method of testing the amount of proteination, chelation or complexing of the mineral to the organic ligand in field samples.

The first step in the utilisation of any nutrient has to be digestion and then absorption. Chelation of the metal trace element to the animal is not new or unnatural. Most trace elements are metals and must be present in a soluble ionised form at the site of absorption. This is greatly assisted by acidity which is naturally at its highest in the stomach and the upper part of the small intestine. As the gut contents become more alkaline, the capacity for absorption will reduce with increasing distance from the stomach. When the trace elements are ionised and in solution, they are 'chelated' by specialised 'carrier' proteins in the gut wall. They pass through the wall by a series of transfers and into the circulation.

CONTEXT   www.contextbookshop.com

Organic Minerals

Chelates

Alternatively, the metal trace mineral can be presented to the gut as a chelate in a peptide conformation. This is then absorbed as a simple molecule. This pathway relies on the peptide surviving the digestive system intact until absorption in the small intestine. There is a probability that in some cases the peptide will reach the circulation unchanged and then behave differently to conventional inorganic trace element sources.

Transitional trace minerals can be chelated because of their ionic form enabling them to form co-ordinate covalent bonds with organic ligands. There are a number of commercially available chelates of iron, copper, manganese, zinc, and cobalt. Selenium cannot be chelated in the same manner but organic forms of selenium are commercially available as selenized yeast. In this form, selenium is present within the sulphur-containing amino acids displacing the sulphur.

Literature demonstrates superiority, often a large one, in the ability of the chelated trace element to reach and lift plasma and body tissue concentrations. It is not what is eaten but what the tissues utilise effectively. Larger peptides will be too large to be absorbed through the intestinal wall and so there has to be some digestion of these complexes prior to absorption. The 'chelated' trace element could be exposed prior to arrival at the absorption sites.

Organic trace minerals can provide an alternative source for animals, where the existing supplementation with inorganic sources is imposing some kind of limit on commercially important parameters such as growth, reproduction and/or immune response, health. Organic minerals can reduce or eliminate interferences from other minerals and so also have a sparing effect. Stress and disease put high demands on the body and could affect absorption characteristics.

Organic minerals have shown specific target tissue responses and so might be desirable after a long depletion period and a target trace mineral response is required. E.g hoof condition. If organic minerals are more effectively absorbed and or less is required for response then a reduction in environmental pollution may be an effective criterion for use.

The quality of the starting material, the type of mineral-ligand binding and the stability constant of the resulting compound will impact the quality of organic trace mineral products. Suppliers must be prepared to show that their products can consistently provide a measurable benefit in terms of performance and cost.

Water Quality

G1

Recommended Maximum Concentrations

Recommended maximum concentrations for selected chemicals and minerals in livestock drinking water.

Alkalinity (as CaCO$_3$)   500 mg/l
Alkalinity levels above 500 mg/L can have a laxative effect. Lower levels may have a laxative effect if sulphate is present in the water.

Aluminum (Al)   0.5 mg/l
Upper limit guideline for cattle.

Calcium (Ca)   700 mg/l
Guideline value when magnesium is present

Calcium (Ca)   1000 mg/l
Guideline value when magnesium is absent

Chloride (Cl)   1000 mg/l
>250ppm can cause brackish taste that may result in low water intake. Reduced growth in immature chickens, but effect largely overcome by adding Na and K. Humans <250mg/L   Veal target <5ppm.

Chromium (Cr)   1.0 mg/l
Guideline max for cattle 0.1ppm (NRC), 0.05ppm (U.S.EPA)

Cobalt (Co)   1.0 mg/l
Guideline value. Cobalt is an essential trace element; toxicity symptoms will likely not become apparent until levels an order of magnitude higher than the recommended level is reached.

Copper (Cu) - cattle   1.0 mg/l
   - sheep   0.5 mg/l
   - pigs and poultry   1.0 mg/l

Copper is essential to animal health and is often a feed additive. Revise levels downwards if supplements are given or feed is high in copper.
(0.5mg/l recommended) High levels produce a bitter flavour.

Fluoride (F)   2 mg/l
Guideline value, but mottling of teeth may occur at this level. If fluoride is included in feed, concentration should not exceed 1 mg/L.

Hardness (as CaCO$_3$)

0-60 mg/l	soft
61-120 mg/l	moderately hard
121-180 mg/l	hard
>180 mg/l	very hard

Hardness has no effect on water safety, but can result in the accumulation of scale in water delivery pipes. The scale mainly consists of magnesium, manganese, iron and calcium carbonates. Water with less than 120 mg/L as CaCO$_3$ is ideal.

Iodide (I)   50 mg /day
Reduced reproduction in sheep, 2,500 mg/L no effect on pigs, 625-5,000 mg/L caused reduced egg production, egg size, and hatchability in laying hens.

CONTEXT   www.contextbookshop.com

Water Quality

G2

Recommended Maximum Concentrations

Iron (Fe)   0.3 mg/l

No toxicity guideline established. Veal calves will have increased colouration of meat even at iron levels as low as 0.1 mg/l. This level can also give milk an oxidised flavour. Iron can present problems in restricted flow drinking water lines where iron precipitation may plug the line. It can also present problems when water is disinfected and can encourage bacterial slime growth in water supply lines.

Lead (Pb)   0.1 mg/l

Guideline value. Lead is cumulative and problems may begin at 0.05mg/l. Chronic lead poisoning may occur at levels of 0.5 to 1.0 mg/L. NRC –cattle guide maximum -0.015mg/l

Magnesium (Mg)   150 mg/l

Reduced growth and bone mineralisation in immature chickens. An upper limit of 125 has been suggested for dairy cows and 30 for veal calves. Magnesium forms part of the hardness in water.

Manganese (Mn)   0.05 mg/l

No toxicity guideline established. Manganese together with iron will discolour fixtures. Manganese and iron can present problems in restricted flow devices in drinking water lines where manganese precipitation may plug the line. Manganese will also present problems when the water is to be disinfected. Can also have an adverse effect on taste. Target 0.02mg/l for veal calves.

Molybdenum (Mo)   0.25 mg/l

Guideline value. An essential element, but it is toxic (linked to intake of copper sulphate). Cu:Mo ratio of 2:1 will prevent poisoning. Sheep, swine and poultry are more tolerant than cattle to poisoning.

Molybdenum (total)   0.05 mg/l

Maximum Criterion. In Canada in British Columbia they maintain a 10 times lower value for molybdenum.

Nitrate (NO_3)   0-100 mg/l

Guideline value. Nitrate may impair the oxygen-carrying capacity of the blood by nitrite reducing haemoglobin to methemoglobin. Ruminants are more sensitive than monogastrics because of the ability of the rumen microbes to reduce nitrate to the more toxic nitrite. Often indicates bacterial contamination or contamination from runoff water from land that has received heavy application of fertiliser. Reduced gains, milk production, reproductive problems. Take care that comparing analyses are the same. (I.e)10mg/l nitrate equivalent to 2.3mg/l nitrate-N. Human max. Nitrate-N is 10mg/l

Nitrite (NO_2) - typical   10 mg/l
         - veal target   0-0.1 mg/l

Guideline value. Is more toxic than nitrate. Nitrite may impair the oxygen-carrying capacity of the blood by reducing haemoglobin to methemoglobin. Animal suffocates. >4mg/l maybe toxic to cattle. Nitrite-N human max 1mg/l (3.3mg/l Nitrite).

pH   6.5-8.5 pH units

Guideline values. If pH is lower than 5.5, acidosis and reduced feed intake may occur in cattle, but is unlikely to have an effect on pigs. Chlorination efficiency is reduced at high pH. A low pH may cause precipitation of some antibacterial agents delivered through the water system (for example sulphonamides). Veal calves target: 6.8-7.3.

CONTEXT   www.contextbookshop.com

Water Quality

G3

Recommended Maximum Concentrations

Phosphate (total P)   1 mg/l

Guideline value. In surface waters, phosphate is typically a limiting plant nutrient. Human max. recommended is 0.1mg/l.

Potassium (K)   10 mg/l

As chloride contributes to TDS test. Can affect taste and corrosivity.

Selenium (Se)   0.05 mg/l

Guideline value. An essential element, but at high levels can be toxic.

Sodium (Na)   200 mg/l

Contributes to salinity and TDS. Can affect taste and corrosivity. Target veal 20mg/l.

Sulphate (SO_4)
- typical   1,000 mg/l
- calves   500 mg/l
- humans   500 mg/l
- veal calves   3 mg/l
- poultry   30-50 mg/l

At >500mg/L the form of S is important. Sulphate interacts with copper metabolism in most animals. High sulphate water consumption often requires changes to the mineral mix that one needs to give to the animals. This has two components, increasing the copper, and decreasing some other minerals. >750mg/L has caused diarrhoea in pigs.

Sulphide (H_2S)   <1.0 mg/l

This is not a toxicity guideline, but a taste and smell advisory. High levels may increase problems with anaemia and those related to Cu, Se and Vit E nutrition. Levels above 25 mg/L are required to cause decreased growth in chickens.

Zinc (Zn)   5-25 mg/l

Guideline value essential element for livestock, but at high levels it can exert toxicity. The lowest recorded effect was at 20 mg/L where the rumen microbes in cattle were affected (decreased digestion of cellulose)

Information from CCREM 1987, Australian Water Quality Criteria 1974, NAS 1974, CAST 1974, BCMOELP 1994, NRC 1974, NRC 1998, NRC 2001, Manitoba Agriculture 1992, EPA 1996, CPAQ 1999, Agriculture and Agri-food Canada (2000).

Water quality standard listings of concentrations (as above) would be useful expressed as 'total maximum daily load'. These present a dose risk more accurately and would enable interpreters to consider a specific element acquired from all sources, e.g.. feed and water. For example: When forages, especially grasses, are stressed by drought, they can accumulate nitrates. A high level of nitrogen fertilisation also is a factor in many cases of nitrate poisoning.

No definite guidelines for presence of microbes in livestock drinking water sources. If pollution is from human wastes, faecal coliforms should exceed faecal streptococcus by several times. If pollution is from an animal source, streptococcus should exceed coliform in refrigerated samples run soon after taking.

CONTEXT   www.contextbookshop.com

Water Requirements

Body Needs

Basic Metabolic Functions
Basic metabolic functions need daily water intake to replace that which is continuously used or eliminated i.e. transporting nutrients (blood volume), excretion of waste products (urine and faeces), digestion of feed, maintenance of osmotic pressure, lubrication of joints and eyes, exchange of CO_2 with oxygen in the lungs, regulation of body temperature - especially heat release by the lungs and urine.

During Periods of Water Loss or Restriction
During period of water loss (e.g. scours, hot or humid weather) or water restriction (e.g. ice, dirty water, inadequate supply) reductions in body fluid negatively impacts metabolism and feed intake. Even mild dehydration (1-5% loss of body weight from water loss), with symptoms not visible to the human eye, reduces metabolic efficiency and impairs ability to regulate body heat (ear and leg extremities feel cool to the touch).

What Effects Water Intake?
Ambient temperature, scours, water temperature, feed intake, incorrect feeding.

Where Does Water Come From?
An animal obtains its water from drinking water, water present in its food and metabolic water.

Producers should be aware of their water quality and ensure that water sources used for the animals are fresh, clean and free of contaminants. At least, yearly water analysis is suggested for all water systems and particular attention after snow has melted, heavy rainfall and excessive run-off situations. Particular attention should be paid to well water sources.

Water Consumption Guidelines	(Litres/day)
Cattle - beef	26-66
- dairy	28-110
Cats	0.1-0.3
Chickens	0.2-0.4
Dogs	0.4-2.0
Goats	3.0-15
Horses	30-45
Pigs	11-19
Rabbits	0.15
Sheep	4-12
Turkeys	0.4-0.6

<u>Notes</u>
Based on mature, average weight animals in temperate climate).

Aluminium

1a

| Al | Atomic Wt 26.98 | Atomic No 13 | Minor Mineral |

Introduction
- The most abundant metallic element in the Earth's crust and third most abundant of all elements in the Earths crust, after silicon and oxygen
- A silvery-white metal, very light in weight, yet relatively strong. It is ductile
- A reactive metal that does not occur in metallic state in nature
- Main source is bauxite (rock composed of hydrated aluminium oxides)
- Name Aluminium comes from 'alumen', the Latin name for alum
- Recycling aluminium products is an important source of metal in many countries
- Used in the manufacture of: paint, dyes, catalysts, disinfectants, cosmetics, paper making, refining operations, tanning, fabric painting, food additives, coagulants, abrasives, vanilla powders, defoaming agents, bleaching agent for flour and explosives
- Jet aircraft fuel contains high levels (>0.1ppm). Urban air tends to contain 2 to 3 times that of rural air
- Al is found in the body but there is no evidence that it is required
- Usually found in the bones and soft tissues

Key Natural Sources
- Levels highly variable in foods of plant origin probably a result of ease of soil contamination
- Typical levels are:
  Soil   5-30%
  Grass and clover 10-50ppm
  Barley   3-60ppm
- High levels have been reported in:
  Oat Silage   1000ppm
  Clover   >3000ppm
  Mixed pasture   8000ppm
- Soil and water contamination or presence of contaminated supplements, (some phosphorous sources) may increase levels in diet. Al is predominant cation in acid-mineral soils with pH 5 or less

Function
- It may have a relationship with carbohydrate metabolism and calcium/vitamin D_3 usage
- Not conclusive that it has an essential function

Absorption
- Soluble salts are absorbed better than plant material. Aluminium sulphate best absorbed

Metabolism
- The gut wall provides a barrier with only 1% of ingested material absorbed
- Al appears to accumulate in hair as a result of stress in dogs

Excretion
- Via the faeces and urine

Requirement /Allowances
- No evidence that it is essential to animals

Adequate Status

Species	Liver	Kidney	Blood	Milk
Ruminant	<1-5	2.3-6.0	0.15-0.97	
Dogs	0.015-1.0	0.06-0.08	0.02	1.0-3.0
Units	ppm WW	ppm WW	ppm WW	ppm DM

CONTEXT www.contextbookshop.com

Aluminium

| Al | Atomic Wt 26.98 | Atomic No 13 | Minor Mineral |

Toxicity

General
- A neurotoxic metal
- May damage the kidneys and enzyme production
- May inhibit glucose absorption in the gut, decrease liver glycogen
- Can inhibit formation of bone phosphates
- Toxicity is enhanced by deficiency of Cu, Fe, Mg, Zn

Ruminants
- Modifies cholinergic neuro-transmission
- Neurological disorders – neurotoxicity may be modulated by the levels of tissue and cytoplasmic calcium
- May cause gastrointestinal irritation at very high levels
- 2000ppm Al as chloride or 4000ppm as sulphate lowers serum Mg
- Reduces growth and feed conversion

Poultry
- Reduced plasma P
- Reduced feed intake, weight gain and egg production
- Increased incidence of perosis
- Young chicks, (<2 weeks age) most susceptible

Maximum Dietary Tolerable Level (ppm DM)

Ruminants	1000	Rabbits	200
Pigs	200	Horses	200
Poultry	200		

Interrelationships

Antagonists
- Al affects Ca, P, Mg and F availability High Al reduces serum Mg levels resulting in grass tetany or hypomagnesaemia in ruminants
- Al replaces Ca during hypocalcaemia and interferes with Mg and P metabolism and parathyroid function
- Al forms insoluble complexes with P making it unavailable to plants
- Al reduces P absorption and increases Ca excretion
- Al reduces F absorption from the gut
- High Al levels increase liver Zn and Fe

Synergy
- Vitamin B_6 and C supplements have been found useful in preventing Al build-up in hair of dogs

Main supplements
- No supplementation necessary as levels in natural feedstuffs meet requirements
- Clays such as kaolin and bentonite, used as pellet binders may be a source

Arsenic

| As | Atomic Wt 74.92 | Atomic No 33 | Minor Mineral |

Introduction
- Name 'arsenic' comes from Greek word 'arsenikon' which means orpiment, a bright yellow mineral of arsenic sulphide
- Occurs in two distinct forms, one a brittle, grey metal and the other a yellow, non-metallic form
- Elemental form has few uses, nearly all applications are as salts or oxides
- Has been shown to be an essential trace element for rats, pigs, goats, humans when highly purified and refined diets are fed in low metal environments
- Better known as a cumulative poison under conditions of excess
- As poisoning is not common due to discontinued use of most sources (insecticides, herbicides, defoliants, sheep dips, rodenticides etc.)

Key Natural Sources
- Abundance in earth crust 1.5-2ppm
- Found in foods, water and environment
- Most common form is arsenopyrite, a compound of iron, arsenic and sulphur
- Most As obtained from byproducts in the treatment of gold, silver, copper and other metal ores
- Soil levels determined by industrial pollution, discarded cans of pesticides and erosion. Mining and smelter regions –soils up to 2500ppm As
- Well water near mines may have contamination
- Has been used in oil drilling, weedkillers, cattle and sheep dips, insecticides, wood preservatives and anthelmintics. Ashes from burnt As treated wood may contain high levels. Most uses have stopped due to the poisonous nature of arsenic compounds
- Marine sources can be high (contamination) – fishmeals 2-20ppm, seaweed- 89ppm and very available
- Normal soil average - 5ppm dry weight
- Not readily absorbed by plants from soil
- Normal forage average -0.25ppm dry weight
- Most feeds- <0.5ppm FW

Function
- Potentially can replace other metals in metallo-enzyme complexes
- Forms various metabolites from methionine, including cystine, polamines and taurine
- May be bacteriocidal and bacteriostatic to gastro-intestinal microbes providing some growth promoting effects at low levels

Benefits
- Very very low level thought to be needed for growth and for healthy nervous system

Absorption
- Organic and inorganic forms are equally well absorbed
- Estimated absorption 5 -70%

Metabolism
- Non absorbed As accumulates (in hens) in order of skeletal muscle, liver, blood, lungs, kidneys, feathers, ovaries
- Highest levels usually in hair, skin, nails form binding to S-proteins
- Rumen reduces absorption, incorporation and excretion

CONTEXT   www.contextbookshop.com

Arsenic

2b

| As | Atomic Wt 74.92 | Atomic No 33 | Minor Mineral |

Excretion
- Excess rapidly excreted from blood and tissues mainly via urine

Requirement /Allowances
Work with goats suggested minimum 50 mcg/kg ration DM/day
Most feed and water are expected to be above this level

Adequate Status

Species	Liver	Kidney	Blood	Hair	Urine
Cattle	0.004-0.4	0.018-0.4	0.03-0.05	0.09-3.0	0.17-0.5
Sheep	0.01-0.2	0.01-0.3	0.01-0.08		
Pigs	0.003-0.2	0.003-0.1	0.01		
Poultry	0.01-0.25	0.01-0.2			
Horses	<0.4	<0.4			
Rabbits	1-2.5				
Dogs	<0.2	<0.2			
Units	ppm WW	ppm WW	ppm WW	ppb DM	ppm WW

Deficiency

General
- Reduced integrity of red blood cells
- Reduced growth
- Impaired fertility
- Reduced milk yield and quality (milk fat)
- Poor conception rates
- Increased abortions
- Reduced viability of offspring
- Reduced birth weight
- Sudden heart failure
- Skin and skeletal lesions

Toxicity

General
- Depends on the concentration and form
- Elemental As is non-toxic. Inorganic arsenicals are more toxic than organic arsenicals. Trivalent As (arsenite) more toxic than arsenate
- Arsenite blocks lipoate dependant enzymes
- At high levels and over prolonged period, it is very toxic
- Acute toxicity occurs quickly with haemorrhages of the intestine and removal of the mucous membranes. (colicky pain, vomiting, diarrhoea, depression, dermatitis)
- Maximum recommended level in drinking water for all livestock and wildlife is 1.0mg/l

Poultry
- Reduced egg production, body weight and feed intake

Pigs
- Rapid breathing, become unsettled, lack co-ordination of their muscles and become blind

Ruminants
- Sheep and cattle do not find arsenic distasteful and may develop a craving for it
- Sudden death, colic, ataxia, partial paralysis, salivation, weakness, rumen atony, dehydration, watery or bloody diarrhoea, depression, loss of appetite, seizures, hypothermia or fever

Arsenic

| As | Atomic Wt 74.92 | Atomic No 33 | Minor Mineral |

Maximum Dietary Tolerable Levels Inorganic ppm (NRC 1980)

Ruminants	50	Rabbits	50
Pigs	50	Horses	50
Poultry	50		

Maximum Dietary Tolerable Levels Organic ppm (NRC 1980)

Ruminants	1000	Rabbits	200
Pigs	200	Horses	200
Poultry	200		

Antagonists

- As antagonistic to I, Se, Hg, Pb
- Arsenic binds to sulfhydryl groups of proteins interfering with their function (e.g. Keratin found in hair and skin).
- Interferes with enzymes that extract energy from food
- Disrupts ATP generation in cellular respiration
- Thought to reduce manganese in bones and tissues

Main Supplements

Not applicable as levels in natural feedstuffs meet requirements
Clays such as kaolin and bentonite, used as pellet binders may be a source

Boron

3a

| B | Atomic Wt 10.81 | Atomic No 5.0 | Minor Mineral |

Introduction
- A semi-metallic element, exhibits some properties of a metal and some of a non-metal
- Occurs in nature as borax, colemanite, boronatrocalcite and boracite
- Earths crust concentration about 10ppm
- Borax may be used to control pH in well drilling muds
- Exists in biological material mainly bound to oxygen
- An essential trace mineral for green algae and higher plants
- Boron has not been shown to perform an essential function in animals

Key Natural Sources
- Common in soils as borate (range 7-80ppm)
- In water as orthoboric acid
- Boron fortified fertilisers are used for certain agricultural crops
- Borax and boric acid have been used as antiseptics
- Boron compounds can be used to make water softeners, soaps and detergents
- Boron nitride, one of the hardest known substances, used for abrasives and cutting tools
- Rich sources: fruit, leafy vegetables, nuts and legumes (10+ppm DM)

Function
- May have effect on some vitamins e.g. D_3 and hormones
- May be involved in bone structure
- May somehow regulate parathyroid hormone action

Benefits
- Influences Ca, P, Mg and cholecalciferol metabolism
- Involved in mineral stasis
- Supplemental B has been used to prevent and cure rheumatoid arthritis

Metabolism
- Distributed throughout the tissues and organs of animals
- Highest concentrations in bones, fingernails, hair and teeth

Adequate Status

Species	Diet	Liver	Plasma	Milk	Urine
Cattle	1-50	0.5-1.5	0.5-4.0	0-1.0	65-107
Units	ppm WW	ppm DM	mg/l WW	ppm WW	mg/hr WW

Deficiency

General
- Rarely seen
- Most likely as a result of an induced deficiency in crops from soils that are over limed (i.e.) too alkaline
- Low Ca/Mg retention
- Reduced growth
- Impaired immune function

Cattle
- Poor conception

Pigs
- Poor conception

CONTEXT www.contextbookshop.com

Boron

3b

| B | Atomic Wt 10.81 | Atomic No 5.0 | Minor Mineral |

Toxicity

General
- Low toxicity to animals when fed orally
- Suggested maximum for livestock drinking water is 5ppm. (Normal drinking water <1ppm)

Cattle
- 150-300 ppm Boron in cattle drinking water for 30 days produces toxic signs
- Inflammation and oedema of legs
- Reduced feed consumption and weight gain
- Reduced hematocrit and haemoglobin levels and plasma P
- Lethargy and occasional diarrhoea, or heamorrhagic gastroenteritis
- Weakness and staggering

Pigs
- May affect serum thyroid hormone concentration
- May affect Inflammatory response
- May affect growth in young pigs

Poultry
- Reduced hatchability of eggs (>300ppm DM)

Maximum Dietary Tolerable Level (ppm DM) (NRC 1980)

Ruminants	150	Rabbits	150
Pigs	150	Horses	150
Poultry	150		

Antagonists
- High levels appear to have adverse effects on P metabolism
- High levels are detrimental to Riboflavin, inducing riboflavinuria

Synergy
- Vitamin D deficiency may be improved
- May help with magnesium deficiency

Main Supplements
Supplementation not recommended / required

CONTEXT   www.contextbookshop.com

Cadmium

4a

Cd | Atomic Wt 112.4 | Atomic No 48 | **Highly Toxic Mineral**

Introduction
- A very soft, silvery-white metallic element
- Name, 'Cadmium' comes from Latin word '*Cadmia*' which means 'calamine'
- Has many chemical similarities to zinc and is used in batteries, photo electric cells, electroplating and soft solders
- Also found in textile and pigment manufacturing and in the chemical industry
- Very toxic metal that is released during the smelting of iron ores and refining steel, so has implications to soil, plants and animals in industrial polluted areas
- Is not an essential micronutrient for animals but acts in the body by replacing other micronutrients in important enzyme complexes and reactions – affecting normal metabolism

Key Natural Sources
- Plants and animal tissue occurrence highly correlated with Zn
- Burning fossil fuels and auto exhausts release Cd
- Cd may be present in phosphate rock; levels in dicalcium phosphate vary with manufacturing site
- Found in commercial fertilisers containing phosphates and is a main source of contamination of soils, containing up to 42 ppm Cd depending on source
- Found in water from galvanised or black polyethylene pipes
- Normal soil levels are 0.1-1.0 ppm. Abnormal levels up to 160ppm (smelter proximity)
- Municipal sewage sludge may have a high Cd content

Cadmium (ppm DM)	Normal	High
Soil	0.03	0.21
Grain	0.01-0.07	3.6-4.0
Grass	0.8-1.7	
Corn Silage	<0.02-1.27	
Oats	0.03	0.27
Wheat	0.01	
Clover	0.1	1.38
Vegetables, nuts, fruits	0.04-0.08	

Notes
Plant uptake: Uptake varies with plant species, distance from highways and smelter regions or where urban sludge has been applied (up to 10ppm DM recorded)

Function
- Non essential

Benefits
- Storage of copper in the livers of sheep may be reduced as cadmium intake increases
- Affects copper concentrations in tissues of hens and eggs
- Can stimulate plasma glutathione peroxidase activity if dietary selenium is adequate but not excessive

Absorption
- Estimated 2-8% dietary Cd is absorbed

Cadmium

4b

| Cd | Atomic Wt 112.4 | Atomic No 48 | Highly Toxic Mineral |

Metabolism
- Cadmium is accumulated in the liver, formed into a cadmium-metallothionein complex and then circulated in the blood to the kidneys, where its accumulation can cause renal damage
- Accumulates in the kidney as a function of age
- Accumulated in offal of ruminants in high Cd herbage
- Does not accumulate in muscle or bone
- Hair levels increase with age
- Hair levels increase in winter, as do dormant forage levels
- Will cross placental barrier to foetus at very high exposure levels

Excretion
- Negligible until acute toxicity
- Tissue retention longer than Pb or Hg
- Little deposited in eggs, milk, foetus

Adequate Status

Species	Liver	Kidney	Blood	Hair	Urine
Cattle	0.02-1.0	0.05-1.5	0.001-0.04	0.04-0.6	<0.15
Sheep	0.02-1.4	0.06-0.48	0.004-0.2	0.55-1.22	0.01-0.03
Pigs	0.04-0.5	0.25-0.99			
Poultry	0.04-0.5	0.02-1.5			
Horses	0.01-5.0	0.05-10	0.0003-0.02	<0.1-0.6	<0.006-0.012
Rabbits	0.3-1.0	3.6-20			
Dogs	0.037	0.12-0.18		0.1-0.9	
Units	ppm WW	ppm WW	ppm WW	ppb DM	ppm DM

Notes
The adequate status level for milk from cattle is 0.005-0.2 ppb dw

Deficiency

General
- Little evidence that Cd acts as an essential element

Goats
- Impaired growth, reduced milk production, shortened life span, unthrifty kids

Toxicity

General
- Toxicity results in kidney damage. Renal and intestinal epithelial destruction

Pigs
- 83ppm dietary Cd reduces growth rate, feed consumption
- May lead to proteinuria and formation of Ca phosphate crystals in the kidney. (Cd is more toxic if dietary Ca is low)
- Testicular degeneration can be produced

Ruminants
30ppm dietary Cd has produced:
- Anaemia
- Abortions
- Still births
- Congenital defects
- Impaired growth rate
- Impaired milk production
- Hypertension
- Sodium retention
- Reduced immune response
- Retarded testicular development
- Enlarged joints
- Scaly skin
- Increased mortality

Cadmium

4c

| Cd | Atomic Wt 112.4 | Atomic No 48 | Highly Toxic Mineral |

Toxicity continued

Poultry
- Egg production and egg weights are reduced. Gizzards show epithelial degeneration. (toxic symptoms dependant on dietary level)
- Growth retardation in turkey poults

Rabbits
- 300ppm dietary Cd is toxic in 19-44 weeks

Maximum Dietary Tolerable Level (ppm DM)

Ruminants	0.5	Rabbits	0.5
Pigs	0.5	Horses	0.5
Poultry	0.5		

Notes
Based on human food residue considerations

Interrelationships
- Interacts with zinc, selenium, iron, copper, cobalt, manganese, ascorbic acid

Antagonists
- Cd interferes with Vitamin A metabolism. The transportation across the gut wall may compete with that for zinc
- Cadmium binds to metallothionein thereby competitively decreasing copper absorption and to a lesser degree absorption of zinc
- Liver and Kidney zinc levels increase with increasing dietary Cd levels
- Fe and Cu levels in liver, kidney and blood decrease with increasing dietary Cd levels
- Coccidiosis in poultry increases kidney Cd accumulation

Synergy
- Selenium reduces Cd accumulation and toxicity
- Excess Cd reduces toxic effects of Pb
- Low dietary Ca and/or Fe allows increased Cd absorption

Main Supplements
Cadmium is listed as an undesirable substance with maximum feedstuff contents dictated by EEC feedstuff legislation.

Calcium

Ca | Atomic Wt 40.08 | Atomic No 20 | Macro Mineral

Introduction
- Most abundant mineral element in the animal body
- About 2% of body weight and over 45% of the total minerals in the body
- 98-99% of calcium occurs in the skeleton and usually found along side phosphorous
- Provides hardness and structural strength to the bones and teeth
- 1-2% of calcium is found in body fluids, involved in nerve and muscle function, enzymes, blood clotting
- Pure Calcium is a soft, silvery white metal found most widely in rocks as chalk, limestone and marble, comprises 3.5% of earths crust
- Name 'Calcium' comes from the latin word 'calcis' which means 'lime'
- Found naturally, only in compounds, usually in the form of carbonates, oxides and sulphates
- Calcite, from greek word *'chalix'* for lime, most common basic chemical, produced from calcium carbonate rocks has been used for thousands of years for construction. Also used for chemical and industrial uses. e.g. steel manufacturing, power plant smokestacks to remove sulphur from the emissions, mining, paper and paper pulp production, water treatment and purification, waste water treatment, road construction
- Lime produced in many countries including United Kingdom, USA, Canada, Mexico, Belgium, Brazil, China, France, Germany, Italy, Japan, Poland, Romania
- Lime can be recycled. e.g. Paper companies recycle large volumes of the lime they use

Key Natural Sources
- Levels of crops and forages dependent on soil factors, plant species, stage of maturity, yield, crop management, climate and soil pH
- Forages
- Milk
- Leafy green vegetables
- Nuts
- Root vegetables, Legumes
- Fish meal
- Lucerne/alfalfa pellets
- Cereal by-products

Function
- Muscle contraction
- Enzyme activation
- Controls heartbeat (An increase in calcium causes the heart to beat faster and a reduction will make in beat slower)
- Transmission of nerve pulses
- Hormone secretion
- Blood clotting (needed for pro-thrombin to form thrombin)
- Cell wall permeability/ structure
- Bone and tooth formation

Benefits
- Milk Production
- Egg production (Shell structure calcium carbonate)

Absorption
- Both active and passive
- PTH, Calcitonin and active form of Vitamin D $(1,25(OH)_2$ Vitamin D_3 control calcium homeostasis
- Absorption is throughout most of the intestinal tract; duodenum and jejunum are most active absorptive sites
- Typically ingested calcium absorption is 30%-50%
- Absorption is dependant on availability from feedstuffs and inorganic supplements at contact point with absorbing membranes

Absorption increased by
- Feed source (milk 90% cf <50% from dry feed sources)
- Body needs
- Protein
- Vitamin D
- Lactose
- Acidogenic rations
- Low anion:cation ratio
- Phytase in ration increases the availability of calcium from concentrates
- Vitamin C (chicks)

CONTEXT

Calcium

| Ca | Atomic Wt 40.08 | Atomic No 20 | Macro Mineral |

Absorption decreased by
- High dietary calcium
- An imbalance with Phosphorous
- High fat levels
- Vitamin D deficiency
- Age
- High pH (alkaline intestinal conditions)
- Oxalic acid (1% can reduce absorption by 1/3 in horses)
- Phytic acid (In non-ruminant animals, phytate reduces the availability of calcium from concentrates)
- High fluorine, sulphur

Metabolism
- Process is under hormonal control monitored by parathyroid hormone
- The body tries to maintain a constant level of extracellular calcium
- Low plasma calcium increases intestinal absorption and reduces urinal Ca loss.
- If calcium levels continue to fall below requirements, calcium is reabsorbed from bone to ensure there is sufficient extracellular calcium. This is not a rapid response. It is mobilised most rapidly from the jaw
- Long periods of deficiency lead to a weakening of bone structure causing rickets in young and osteomalacia in adults
- When the calcium loss exceeds intake quickly, acute hypocalcaemia occurs, a condition often found in dairy cows. Hypocalcaemia can result in loss of nervous and muscle function with the clinical condition known as milk fever
- Low Mg reduces Ca mobilisation into the blood
- Ca is transferred to the egg and through the placenta

Excretion
- Controlled by hormones
- Faeces are primary path of Ca excretion
- Urinary loss is minimal and a little is lost from sweat
- Urinary loss may by higher when high levels of Ca in feed. (more so hindgut fermenters)

Requirement /Allowances (g/kg DM)

Rums	NRC	Pigs	NRC	Poultry	NRC	Others	NRC
Calf (a)	5	Creep	9	Chick	9-10	Dog	6-10
Dairy	6-8	Weaner	8	Broiler	9	Cat	8
Beef	1-5	Grower	7	Breeder	32.5	Horse	2.4-6.8
Heifer	4.1-3.7	Finisher	6	Layer	32.5	Fish	10
Sheep	2-8.2	Sow/Boar	7.5	Turkey	5-12	Rabbits	4-7.5

Rums	Typical	Pigs	Typical	Poultry	Typical	Others	Typical
Calf (b)	7 (8)	Creep	9	Chick	10	Dog	6-10
Dairy (c)	7-8	Weaner	9	Broiler	10	Cat	6-10
Beef	3.5-6	Grower	8.5	Breeder	32.5	Horse	4-7.5
Heifer	3-5	Finisher	8.5	Layer	36	Fish	
Sheep	3.5-7	Sow/Boar	8-9	Turkey	8-13	Rabbits	

Notes
(a) Calf on milk replacer 7 g/kg DM (NRC)
(b) Calf on milk replacer 8 g/kg DM
(c) Dry cows 3 g/kg DM

- Levels influenced by age, genetics, level of productivity, physiological state of animal, variability of ingredient nutrients, nutrient availability, nutrient interactions, stress, adequacy of vitamin D intake and liver and kidney integrity.

Calcium

5c

| Ca | Atomic Wt 40.08 | Atomic No 20 | Macro Mineral |

Adequate Status

Species	Liver	Kidney	Serum	Milk	Egg Shell
Cattle	30-200	45-200	8-11	1200-1400	
Sheep	38-80	60-140	11-13		
Pigs	34-65	60-125	9-12	2000-2400	
Poultry		200-600	10-40 (a)		38-44
Horses	40-60	50-250	10-13	0.65-1.2	
Rabbits			10-20		
Dogs	33-250	50-200	9-12		
Cats			7-11		
Units	ppm WW	ppm WW	mg/dl DM	mg/l DM	% DM

Notes
(a) Serum level for layers should be between 20-40 mg/dl and broilers 10-16 mg/dl.

- Serum calcium level depends on analytical technique used
- Pigmented/coloured hair contains more calcium than white hair. Levels do not correlate with dietary intake
- Elevated hair Ca may result from bone mobilisation
- Normal eggshell calcium:38-44%

Deficiency

General
- Young animals: Rickets : enlarged joints of long bones, soft bones bend and animal walks in distorted way (Vitamin D link)
- Adults: Osteomalacia : adult rickets (softening of bone)
- Osteoporosis: defective formation of protein matrix on which bone mineral is laid down. Bones become more susceptible to fracture
- Reduced and depraved appetite
- Growth retardation
- Loss of weight
- Reduced fertility

Pigs
- Young growing animals, sows in late gestation and lactation most susceptible
- Posterior paralysis (sows), hump back syndrome

Poultry
- Thin shells, poor hatchability, reduced egg production, poor egg quality, leg problems

Ruminants
- Parturient paresis; Hypocalcaemia; Milk fever (Mg,P link)
- Occurs from 1-2 days prior to calving but most common at start of lactation due temporary imbalance of Ca availability and high Ca demand at the onset of lactation
- Relates to previous Ca intake and malfunction of hormone form of Vit D and PTH
- Older animals more susceptible
- Reduced milk production

Sheep
- Severe growth stunting, gross dental abnormalities

Dogs
- Tetany and convulsions may result

Horses
- Bone deformities

CONTEXT   www.contextbookshop.com

Calcium

5d

| Ca | Atomic Wt 40.08 | Atomic No 20 | Macro Mineral |

Toxicity

General
- A relatively safe nutrient not associated with specific toxicity
- Not common as intestine regulates it
- May reduce feed consumption, weight gain and delay sexual maturity
- Bone and joint abnormalities
- May reduce fat digestibility
- Kidney stones
- Calcinosis: deposition of Ca salts in soft tissues e.g. tendon (particularly from ingestion of solanaceous plants)

Poultry
- Impairs productivity, growth and reproduction (Chicks with double the needed calcium will suffer growth retardation due to hypercalcaemia)

Maximum Dietary Tolerable Level (g/kg DM)

Ruminants	20	Rabbits	20
Pigs	10	Dogs	25
Poultry (a)	12	Horses	0.5

Notes
(a) laying poultry is 40g/kg.
*maximum levels assume adequate dietary P (Ca:P ratio important) and depending on age and production status

Main Supplements

Source	Element %	Relative Bio-avail.
Bone charcoal (a)		
Bone meal (steamed) (a)	23-37	high
Calcium carbonate	38.5	medium
Calcium chloride	36	high
Calcium citrate		
Calcium gluconate		
Calcium hydroxide		
Calcium lactate		
Calcium oxide		
Calcium sulphate		
Dicalcium phosphate (a)	23.2	high
Dolmitic limestone	22.3	medium
Kelp		
Limestone (ground)	33-38	medium
Monocalcium phosphate (a)	16.2	high
Oyster shells	38	
Soft rock phosphate (a)	15-18	low
Trical phosphate (a)	31-34	

Notes
(a) sources of phosphorous but also supplying calcium

- Mineral supplements contain calcium, and is usually more available than forages and feedstuffs

CONTEXT   www.contextbookshop.com

Calcium

5e

| Ca | Atomic Wt 54.94 | Atomic No 25 | Macro Mineral |

Interrelationships
- Ca, P and Mg directly interrelate

Antagonists
- Excess Ca may reduce the absorption of F, Mg, Mn, P, Zn, Cu, Pb, Cd, Fe, I and possibly other elements
- Ca increases need for P, Vitamin D
- Ca in oxalate form (alfalfa) is unavailable
- Mg deficiency reduces Ca mobilisation into blood
- Ca absorption is reduced by excess Mg, P, S
- Layers: calcium deficiency exacerbated by high levels of dietary chloride (0.4-0.5%)
- Hypocalcaemic cows are more susceptible to Al toxicity

Synergy
- Ca and Mg work together for the health of the heart and circulation
- Ca and P work together to maintain the health of bones and teeth
- Vitamin D is required for calcium absorption. Dietary levels important

Calcium : Phosphorous Ratio
- A dietary Ca:P ratio between 1:1 and 2:1 is assumed ideal for growth and bone formation. This is approximately the ratio in bone
- Vitamin D status affects ratio tolerance
- Pigs 1:1 to 2:1, (max. 1.3:1 in low P diets, 2:1 in high P diets)
- Horses 1:1 to 2:1
- Cattle >1:1 to maximum 7:1
- Sheep 2:1 max (4:1)
- Chickens 1.3-2:1 (laying hens 5-6.5:1)

Calcium

| Ca | Atomic Wt 40.08 | Atomic No 20 | Macro Mineral |

Feed Name	g/kg DM
Alfalfa meal	15.5
Bakery waste	1.4
Barley grain	0.5
Bean field	1.2
Blood meal	3.2
Brewers grains	3.6
Buckwheat grain	1.1
Buttermilk dehyd. (cattle)	14.4
Casein dehyd. (cattle)	6.7
Cassava tubers dehyd.	2.1
Citrus pulp dried	18.2
Copra meal	2.2
Cottonseed whole	1.5
Cottonseed meal	2.3
Distillers grains - wheat	1.7
Distillers grains maize	1.6
Distillers grains - barley	1.7
Fishmeal (Sth Am)	41.5
Grass bluegrass	3.3
Grass alfalfa	19.6
Grass bermuda	5.3
Grass clover	17.1
Grass extensive	4.8
Grass kikuyu	3.3
Grass timothy	3.3
Groundnut ext	1.7
Hay alfalfa	17.8
Hay bluegrass	3.3
Hay clover	15.3
Hay eragostus	2.3
Hominy feed	0.6
Linseed meal (mech ext)	4.3
Maize bran	0.4
Maize germ ext(sol)	0.4
Maize gluten 20	3.3
Maize gluten 60	0.4
Maize grain	0.2
Malt culms	2.3
Milk (cattle-dehyd)	9.5
Milk skimmed	13.6
Millet grain	0.3
Molasses - beet	3.0

Feed Name	g/kg DM
Molasses - cane	10.0
Oat groats	0.8
Oat middlings	1.9
Oatfeed	1.2
Oats grain	0.9
Palm kernel exp.	2.2
Peas	1.1
Potato dried	0.1
Rape ext. (mech.)	7.2
Rice bran	0.9
Rice grain	0.7
Rye grain	0.7
Safflower ext. solv.	3.7
Sesame ext. mech.	21.7
Silage alfalfa	13.3
Silage grass	4.5
Silage maize	1.9
Silage sorghum	3.5
Silage wholecrop	1.9
Sorghum grain	0.4
Soya ext. solv	3.4
Soya flour	4.9
Soya hipro	2.8
Straw barley	4.0
Straw oat	3.3
Straw wheat	3.0
Sugar beet pulp (dehyd.)	6.9
Sugar beet pulp (mol)	7.4
Sunflower ext	3.3
Triticale grain	0.6
Wheat (caustic)	0.5
Wheat bran	1.5
Wheat feed	1.0
Wheat germ ext.	0.6
Wheat grain	0.6
Whey low lactose	17.1
Whey (cattle dehyd)	9.2
Yeast (brewers dehyd)	1.3
Yeast (torula dehyd)	5.4

CONTEXT www.contextbookshop.com

Chlorine

6a

| Cl | Atomic Wt 35.45 | Atomic No 17 | Macro Mineral |

Introduction
- A greenish-yellow strong smelling gas
- Name 'Chlorine' comes from the Greek word *'khlôros'* (green)
- A widespread element especially in salts
- Body contains approx. 0.11% Cl
- Found in the bodies soft tissues
- Found at high levels in the cerebrospinal fluid
- Chlorine - used primarily in producing polymers that are used in manufacture of plastics, synthetic fibres and synthetic rubber; also used in crude oil refining, for making pesticides; in household bleach, water treatment and sewage treatment
- Hydrochloric acid – used in making synthetic rubber and in cleaning gas and oil wells

Key Natural Sources
- Abundant in igneous rock and 1.9% of seawater
- Mainly exists as sodium chloride
- Plant levels affected by fertiliser application (e.g. potassium chloride)
- Found in many feeds e.g. Lucerne/alfalfa, legumes, barley straw, beet molasses, cane molasses, fishmeal, milk by-products, oats, turnips, vetch

Function
- Major anion of extracellular fluid, >60% percent of the total anion equivalents
- Found in large concentrations within and without the cells of body tissues
- Major electrolyte, with K and Na, in regulation of acid-base balance
- Essential for carbon dioxide transport
- Involved in producing gastric juice and stomach pH
- Component of HCL (hydrochloric acid), important in protein digestion
- HCL assists iron absorption (solubilisation and reduction of ferric to ferrous state)
- Activation of enzymes, e.g. pancreatic amylase
- Involved in respiration and regulation of blood pH
- Maintain osmotic pressure and body fluid balance
- Regulation of water balance

Benefits
- Osmotic balance
- Acid-base balance
- Suppresses microbial growth
- Acidic diets encourage Ca absorption in ruminants

Absorption
- Principally from upper small intestine with no apparent controls
- Absorption is very efficient (90%- feeds, 95-100% salts, gastric juices) provided glucose is available for transport purposes
- Salt is solubilised, releasing the negatively charged chloride ion for absorption
- Ruminants, absorption also occurs in the reticulorumen, abomasum, omasum and duodenum as well as intestines
- Absorption of Cl is by passive diffusion following sodium along an electric gradient 80% of Na + Cl entering GI tract is from internal secretions e.g. saliva, gastric fluids, bile, pancreatic juice

Metabolism
- Gastric secretion is largely hydrochloric acid. Regulation is linked to sodium but it can be conserved independently e.g. reduce excretion at the kidney, and in faeces and milk
- Chloride is passively reabsorbed in the kidney
- Salt appetite is an overriding force for Na and Cl intake in omnivores and herbivores

CONTEXT www.contextbookshop.com

Chlorine

6b

| Cl | Atomic Wt 35.45 | Atomic No 17 | Macro Mineral |

Excretion
- Mainly by urine as salts. Level is directly correlated to the level in the diet
- The kidneys excrete chloride under the control of aldosterone
- Renal excretion of excess sodium is accompanied by excretion of chloride
- Smaller amounts lost in faeces and sweat
- Also lost from vomiting and diarrhoea
- Chloride excretion influenced by bicarbonate ion. (increase plasma bicarbonate, increase excretion of Cl)
- Milk contains about 1-1.5g/l chlorine

Requirement /Allowances (g/kg DM)

Rums	NRC	Pigs	NRC	Poultry	NRC	Others	NRC
Calf (a)		Creep	2.5	Chick	1.5	Dog	0.4
Dairy	2.5-4	Weaner	2	Broiler	1.2-2	Cat	1.9
Beef		Grower	1.5	Breeder	1.3	Horse	
Heifer	1-1.2	Finisher	0.8	Layer	1.3	Fish*	9
Sheep		Sow/Boar	1.2-1.6	Turkey	1.2-1.5	Rabbits*	3

Rums	Typical	Pigs	Typical	Poultry	Typical	Others	Typical
Calf	2(b)	Creep		Chick	1.8	Dog	0.9- 4.5
Dairy	2.5-4	Weaner	3.6	Broiler	1.2-2	Cat	3
Beef	2	Grower	3	Breeder	2.5	Horse	1.8
Heifer	2	Finisher	3	Layer	2	Fish	
Sheep	2.5	Sow/Boar	2.7-3.7	Turkey	1.2	Rabbits	

Notes
(a) Calves on milk replacer 2 g/kg DM (b) Milk replacer 8-10 g/kg DM
* Lactation, pregnancy, and growth affect the requirement for chloride (salt)

- Factors influencing salt requirements: dry vs green forage; levels in water; genetic differences for milk levels, illness or disease (diarrhoea, vomiting, renal losses); resistance to disease and parasitism, geographical location, temperature and humidity, sweating capability of species animal class and physiological status, type of feeds
- Calves: When salt (sodium chloride) is used to meet the Na requirement the Cl requirement is met or exceeded. When sodium bicarbonate or some other Na-containing salt is used to supply sodium additional chloride e.g. potassium chloride will be required

Adequate Status

Species	Liver	Serum	Body	Ruminal fluid
Cattle	1044	95-110	1-1.5	10-30
Sheep		95-110		
Pigs		100-105		
Poultry		152		
Horses		98-109		
Rabbits				
Dogs	1490	103-115		
Units	ppm WW	meq/l DM	g/kg DM	meq/l DM

CONTEXT   www.contextbookshop.com

Chlorine

| Cl | Atomic Wt 35.45 | Atomic No 17 | Macro Mineral |

Deficiency

General
- Alkalosis (excess alkali in the blood) resulting in slow shallow breathing
- Muscle cramps
- Reduced food intake
- Convulsions
- Poor growth
- Reduced feed efficiency (only observed on purified or concentrated diets)

Pigs
- Reduced fertility
- Reduced piglet birthweight

Turkey
- Leg weakness
- Dehydration
- High mortality
- Nervous symptoms

Young Calves
- Early signs:
  - Anorexia and lethargy
  - Mild polydipsia and polyuria
  - Weight loss
  - Severe eye defects
- Later signs:
  - Blood and mucus appear in faeces
  - Cardiovascular depression
  - Constipation

Cattle
- Unlikely in grazing animals
- Can induce deficiency if non Cl salts used as Na supplements
- Signs:
  - Reduced milk yield
  - Poor body weight
  - Rough hair coat
  - Polyuria
  - Fertility disorders
  - Pica
  - Lethargy
  - Cardiovascular depression
  - Mild dehydration or constipation
  - Depraved appetite
  - Licking urine, pipes etc

Chicken
- High mortality
- Nervous signs induced by sudden noise
- Reduce egg production and size
- Increased blood packed cell volume
- Decreased plasma CL and Na levels
- Possible reduced immune response

Dogs
- Fatigue, exhaustion, inability to maintain water balance, decreased water intake, retarded growth, dry skin, loss of hair

Toxicity

General
- Rarely seen unless water limiting. Also see sodium for more info
- High chloride levels without the neutralising cation (e.g.Na) will contribute to acidosis
- Effects depend on age, sex, pregnancy, lactation, exercise, climatic conditions, diet type and moisture content, level of production, access to water and adaptation to intake level
- Chicks are more tolerant than turkey poults. Ducks are more susceptible than chicks
- Excess Cl increases incidence of tibial dyschondroplasia

Maximum Dietary Tolerable Level NaCl (g/kg DM)

Lactating cattle	40	Poultry	20
Non-lactating cattle	90	Chicks	>9
Sheep	90	Hens	>12
Horses	30	Pigs	80
Rabbits	30		

Maximum Dietary Tolerable Level Cl (g/kg DM)

Lactating cattle	24.3	Poultry	12.1
Non-lactating cattle	54.6	Chicks	5.4
Sheep	54.6	Hens	7.2
Horses	18.2	Pigs	48.5
Rabbits	18.2		

www.contextbookshop.com

Chlorine

| Cl | Atomic Wt 35.45 | Atomic No 17 | Macro Mineral |

Interrelationships
- Cl interacts with Na and K

Antagonists
- Increasing dietary Cl increases Plasma K linearly. Aid in potassium conservation
- Pigs- Na:Cl imbalance can upset lysine:arginine ratios due to a change in urea cycle activity

Synergy
- Na and CL must be in balance for optimum performance
- Lactating cows -Increase levels during heat stress and increased levels may help with metabolic acidosis
- Chicken- Soluble sulphate has a sparing effect on Cl requirement
- Up to 0.42% dietary Cl (as salt) has increased egg production when ration contains 3.5% Ca and 0.4%P
- Turkey: high Cl reduces toxic effects of excessive Mg supplementation

Main Supplements
- Na and Cl most often limiting if not supplemented
- Poultry- Salt supplement minimised to reduce moisture level in excreta
- Water may play a key role in supplying chloride for high yielding milk producing animals

Main Supplements

Source	Element %	Relative Bio-avail.
Sodium chloride	60	high
Ammonium chloride	65	high
Potassium chloride	47	high
Magnesium chloride	70	

Chlorine

6e

| Cl | Atomic Wt 35.45 | Atomic No 17 | Macro Mineral |

Feed Name	g/kg DM
Alfalfa meal	0.9
Bakery waste	16.1
Barley grain	1.6
Bean field	0.7
Blood meal	3.0
Brewers grains	1.6
Buckwheat grain	0.5
Buttermilk dehyd. (cattle)	4.3
Casein dehyd. (cattle)	0.4
Cassava tubers dehy	0.6
Citrus pulp dried	0.2
Copra meal	0.7
Cottonseed whole	4.7
Cottonseed meal	0.5
Distillers grains - wheat	3.1
Distillers grains maize	0.8
Distillers grains - barley	3.2
Fishmeal (Sth Am)	14.8
Grass bluegrass	4.0
Grass alfalfa	4.7
Grass bermuda	-
Grass clover	7.7
Grass extensive	7.0
Grass kikuyu	13.2
Grass timothy	5.7
Groundnut ext	3.0
Hay alfalfa	5.3
Hay bluegrass	5.3
Hay Clover	3.2
Hay Eragostus	4.2
Hominy feed	0.6
Linseed meal (mech ext)	0.5
Maize bran	0.8
Maize germ ext (sol)	0.5
Maize gluten 20	2.4
Maize gluten 60	0.8
Maize grain	0.5
Malt culms	2.2
Milk (cattle-dehyd)	9.2
Milk skimmed	11.7
Millet grain	0.2
Molasses - beet	12.5

Feed Name	g/kg DM
Molasses - cane	26.6
Oat groats	0.9
Oat middlings	0.6
Oatfeed	0.4
Oats grain	1.1
Palm kernel exp	1.0
Peas	0.5
Potato dried	2.8
Rape ext. (mech.)	0.7
Rice bran	0.6
Rice grain	0.9
Rye grain	0.3
Safflower ext. solv.	0.8
Sesame ext mech	0.8
Silage alfalfa	4.1
Silage grass	8.5
Silage maize	4.0
Silage sorghum	1.3
Silage wholecrop	3.4
Sorghum grain	1.0
Soya ext.solv	0.5
Soya flour	4.2
Soya hipro	0.5
Straw barley	7.0
Straw Oat	8.3
Straw wheat	3.2
Sugar beet pulp (dehyd)	0.4
Sugar beet pulp (mol)	4.5
Sunflower ext.	1.4
Triticale grain	0.3
Wheat (caustic)	0.7
Wheat bran	0.5
Wheat feed	0.7
Wheat germ ext.	0.8
Wheat grain	0.8
Whey low lactose	11.0
Whey (cattle dehyd)	0.8
Yeast (brewers dehyd)	0.8
Yeast (torula dehyd)	0.2
....	
....	
....	

CONTEXT www.contextbookshop.com

Chromium

| Cr | Atomic Wt 52 | Atomic No 24 | Micro Mineral |

Introduction
- Makes up about 1/3000 of earths crust
- Metallic element present in small amounts in all animal tissues
- An essential element, hormone and involved with vitamins
- Trivalent Cr is the most stable and found in association with nicotinic acid, glutamic acid, glycine, and cysteine, commonly known as glucose tolerance factor (GTF)

Key Natural Sources
- Cr Ore is used in the production of stainless steel
- Chromate is used in oil fields as a drilling aid, production of pigments for ink, paint etc
- Used as a wood preservative (often with As and Cu); ashes from burnt treated wood may contain toxic levels of Cr
- Feed levels are often difficult to ascertain due to difficulty in testing accuracy
- More abundant in soils that feed sources
- Plant uptake and animal absorption are limited
- Processing foods can increase Cr, leaching from stainless steel, particularly under acidic conditions
- Liver meal, pulses, seeds, dark chocolate – highest sources
- Fruits, vegetables, grains – variable
- Meat, poultry, fish, dairy - low

Function
- Component of glucose tolerance factor (GTF) that enhances the effect of insulin
- Stimulates insulin activity, glucose uptake by organs and muscle, glucose metabolism and stimulates the synthesis of proteins
- Enzyme activator involved in the production of energy from carbohydrates, fats and protein
- Involved in nucleic acid metabolism and therefore protein synthesis
- Cellular action of Cr may be involved in regulation of cell growth
- Stimulates fatty acid and cholesterol production in the liver

Benefits
- Improves humoral and possibly cell mediated immune response
- Stabilises nucleic acids e.g. DNA/RNA
- Improves glucose tolerance
- Improves dry matter intake
- Improves growth rates
- Improves carcass lean/ decrease carcass fat
- Reduces egg yolk and serum cholesterol
- Improves milk production
- Increases litter weights and size – piglets
- Reduced morbidity in stressed calves

Absorption
- Prior Cr status and dietary intake affects absorption
- Primarily absorbed in small intestine
- Inorganic sources are poorly absorbed and utilised (i.e.) less than 5%
- Organic sources such as protein, nucleic complexes, yeasts are better absorbed (20%+) and different tissue distribution
- Bioavailability from most feedstuffs is extremely low

Metabolism
- Inorganic chromium is less efficiently utilised than GTF chromium
- Storage not high but most found in liver, spleen, soft tissue and bone
- GTF chromium is stored in higher levels in the liver than the inorganic salt
- Next to the liver the kidney is the most active source of GTF chromium
- Chromium requirements of the body may become more critical at times of stress, malnutrition and blood loss
- Transported in blood via transferrin
- Albumin is an acceptor and transporter of Cr if no transferrin binding sites available.

CONTEXT   www.contextbookshop.com

Chromium

7b

| Cr | Atomic Wt 52 | Atomic No 24 | Micro Mineral |

Excretion
- Mainly via the urine
- Small losses from faeces, sweat, hair

Requirement /Allowances

No specific recommendation as to dietary form and concentration of chromium supplements for cattle, poultry and swine	Research has identified situations in which chromium supplementation might have commercial application. e.g. periods of stress (environmental and metabolic)

Adequate Status

Species	Liver	Kidney	Blood	Milk	Hair
Cattle	0.04-3.8	0.5-6.2	0.006-0.06	0.008-0.25	0.2
Sheep					
Pigs					
Poultry *	0.1-0.4				
Horses					
Rabbits	0.3-1	0.42-1.58			
Dogs					
Units	ppm WW	ppm WW	ppm WW	ppb WW	ppm DM

Notes
* eggs 0.05-0.15 ppm ww

Deficiency

General
- Affects glucose metabolism associated with high blood sugar and the loss of sugar in the urine
- Affects protein and lipid metabolism
- Affects insulin output and activity leading to hyperglycaemia, increased blood cholesterol levels
- Protein synthesis is indirectly affected by chromium
- Reduced growth rate, leads to necrotic liver degeneration and increased mortality

Toxicity
- Signs include scouring, vomiting, dehydration, dermatitis, skin allergy and liver, kidney, central nervous system damage. Irritation of respiratory passages, ulceration and perforation of nasal septum
- Reduced growth rate
- The valency form heavily affects toxicity
- Hexavalent chromium most toxic
- Inorganic chromium is much more toxic than similar amounts of GTF chromium
- Can get Cr- related dermatitis from detergents and bleaches
- General levels in feed are too low to cause toxicity

Maximum Dietary Tolerable Level Chloride (ppm DM)

Ruminants	1000	Rabbits	1000
Pigs	1000	Horses	1000
Poultry	1000		

Maximum Dietary Tolerable Level Oxide (ppm DM)

Ruminants	3000	Rabbits	3000
Pigs	3000	Horses	3000
Poultry	3000		

CONTEXT www.contextbookshop.com

Chromium

7c

| Cr | Atomic Wt 52 | Atomic No 24 | Micro Mineral |

Interrelationships:
Zn, Fe, Mn, Vn

Antagonists
- Absorption of the inorganic form may be affected by zinc, iron, manganese and vanadium
- Antacids, phytate, oxalates reduce absorption
- Animals fed high carbohydrate diets can deplete the supply of GTF chromium
- Cr and Zn are antagonistic

Synergy
- Needs niacin to operate in Glucose tolerant factor
- Vitamin C and aspirin promote absorption

Main Supplements

Source	Element %	Relative Bio-avail.	Comments
Chromium Chloride			
Chromium Oxide			
Chromium Nicotinate			molecular wt 418
Chromium Picolinate			molecular wt 418
Chromium Yeast			
Chromium -L-Methionine			water soluble
Chromium Chelate			molecular wt 300 approx
Chromium Proteinate			not soluble

Notes
The need for supplementation depends on
1. Chromium status of animal
2. Amount of bioavailable Cr in feedstuffs
3. Exposure of animal to environmental and metabolic stresses

- Feeds contain some chromium with plant levels normally 0.1-5.0 ppm and the maximum recorded level 3000ppm
- Water (typical river water) contains 1-10ppb

Chromium

| Cr | Atomic Wt 52 | Atomic No 24 | Micro Mineral |

Feed Name	mg/kg DM	Feed Name	mg/kg DM
Alfalfa meal	-	Molasses - cane	-
Bakery waste	-	Oat groats	-
Barley grain	0.1	Oat middlings	-
Bean field	-	Oatfeed	-
Blood meal	-	Oats grain	-
Brewers grains	-	Palm kernel exp.	-
Buckwheat grain	-	Peas	-
Buttermilk dehyd.(cattle)	-	Potato dried	-
Casein dehyd. (cattle)	-	Rape ext (mech)	-
Cassava tubers dehy	-	Rice bran	-
Citrus pulp dried	-	Rice grain	-
Copra meal	-	Rye grain	-
Cottonseed Whole	-	Safflower ext. solv.	-
Cottonseed meal	-	Sesame ext mech	-
Distillers grains - wheat	-	Silage alfalfa	-
Distillers grains maize	-	Silage grass	0.2
Distillers grains- barley	-	Silage maize	-
Fishmeal (Sth Am)	0.9	Silage sorghum	-
Grass bluegrass	-	Silage wholecrop	-
Grass alfalfa	-	Sorghum grain	-
Grass bermuda	-	Soya ext.solv	-
Grass clover	-	Soya flour	-
Grass extensive	-	Soya hipro	-
Grass kikuyu	-	Straw barley	-
Grass timothy	-	Straw oat	-
Groundnut ext	-	Straw wheat	-
Hay alfalfa	-	Sugar beet pulp (dehyd)	-
Hay bluegrass	-	Sugar beet pulp (mol)	0.6
Hay clover	-	Sunflower ext	-
Hay eragostus	-	Triticale grain	-
Hominy feed	-	Wheat (caustic)	0.2
Linseed meal (mech ext)	-	Wheat bran	-
Maize bran	-	Wheat feed	-
Maize germ ext(sol)	-	Wheat germ ext.	-
Maize gluten 20	-	Wheat grain	0.2
Maize gluten 60	-	Whey low lactose	-
Maize grain	-	Whey (cattle dehyd)	-
Malt culms	-	Yeast (brewers dehyd)	-
Milk (cattle-dehyd)	-	Yeast (torula dehyd)	-
Milk skimmed	-		
Millet grain	-		
Molasses -beet	-		

CONTEXT www.contextbookshop.com

Cobalt

8a

| Co | Atomic Wt 58.93 | Atomic No 27 | Micro Mineral |

Introduction

- Makes up only 0.001-0.002% of earths crust where usually found combined with arsenic and sulphur.
- A bluish-gray, shiny, brittle, magnetic metallic element.
- Closely resembles copper in many of its properties
- Body contains about 0.2ppm cobalt Found throughout the body in organs, tissues and blood, usually combined with Vitamin B_{12}
- An essential constituent of Vitamin B_{12} (cobalamin contains 4.5% cobalt)
- Used in superalloys for jet engines, chemicals (paint driers, catalysts, magnetic coatings, pigments, rechargeable batteries), magnets, and cemented carbides for cutting tools
- A man made isotope of cobalt, cobalt-60, produces gamma rays. This is used for sterilisation of medical supplies and foods, for industrial testing, and to fight cancer
- Main use of cobalt is through its salts
- Ancient civilisations in Egypt and Mesopotamia used cobalt to colour glass a beautiful deep blue
- Name 'Cobalt' comes from the German word 'kobald' which means goblin or evil spirit believed to cause health problems for silver and copper miners
- Principal cobalt producing sources Africa, Canada, Australia and Russia
- With exception of P and Cu, Co is most severe mineral limitation in grazing ruminants in tropical countries

Key Natural Sources

- Soil and plant concentration are highly variable
- Soil variation due to large differences in Co content of basic rock (<2ppm soil, deficient for ruminants)
- Plant levels affected by soil concentration, fertilisation, plant species, stage of maturity, yield, soil drainage, soil pH, climate
- Neutral to acid pH ensures a better plant uptake
- Legumes generally have greater ability to concentrate Co than grasses
- Other minerals, e.g. Mo, affect plant availability
- Plant levels vary from < 0.0005 to >0.5 mg/kg DM
- Liver, kidney tend to have highest levels

Rock Type (mg/kg)	Cobalt Content Mean	Range
Basaltic igneous	50	24-90
Granitic igneous	5	1-15
Shales and clays	20	5-25
Black shales	10	7-100
Limestone	0.1	
Sandstone	0.3	

Function

- Ruminants require daily intake for rumen microbes to manufacture Vitamin B_{12}
- Maintains nervous system integrity
- Cobalt is not an essential mineral for pigs or poultry (monogastrics) as they rely on dietary Vitamin B_{12} and can not synthesise the vitamin from cobalt
- Vitamin B_{12} is involved in metabolism of nucleic acids and proteins, formation of proteins from amino acids, carbohydrate and fat metabolism
- Vitamin B_{12} is involved in energy metabolism of propionate to produce glucose. (High starch rations may need a higher cobalt allowance)
- Vitamin B_{12} promotes red blood cell synthesis (haemoglobin and myoglobin)
- Vitamin B_{12} – essential part of several enzyme systems

Benefits

- Involved in enzyme reactions, immune response and ensuring the health of the nervous system
- Levels of cobalt above that needed for Vitamin B_{12} production (up to 0.35 mg/kg of dietary DM) may improve digestion of poor quality feeds in the rumen. This maybe due to some forage microbes having a higher cobalt requirement or by increasing the total rumen anaerobic bacterial population and increasing lactic acid production

Cobalt

| Co | Atomic Wt 58.93 | Atomic No 27 | Micro Mineral |

Absorption
- Cobalt needs an intrinsic factor (a glycoprotein) secreted in the stomach for absorption
- Non-ruminants (maybe ruminants) absorption shares intestinal mucosal transport with iron
- Cobalt absorption by ruminant is less efficient than simple-stomached animals
- About 3% is converted to Vitamin B_{12} in the rumen (efficiency of conversion inversely related to Co intake)
- Of total vitamin B_{12} produced, only 1-3% is absorbed, mostly in lower portion of small intestine
- Vitamin B_{12} requires carrier compounds to pass through intestinal wall e.g. food proteins

Metabolism
- Secreted in the bile and reabsorbed Vitamin B_{12} is stored in adult ruminant and usually will suffice for several months
- Young animals may be more sensitive to cobalt deficiency e.g. calves from Co deficient cows require supplementation within 1-3 months
- Rumen microorganisms need cobalt and within a few days of a deficiency succinate production increases at the expense of propionate production
- 43% of body Co is stored in muscles and 14% in bone, remainder in other tissues (most in liver and kidney)

Excretion
- Mainly via faeces
- Small amounts lost via urine, sweat and hair

Requirement /Allowances (mg/kg DM)

Rums	NRC	Pigs	NRC	Poultry	NRC	Others	NRC
Calf	0.1	Creep		Chick		Dog	
Dairy	0.11	Weaner		Broiler		Cat	
Beef	0.1	Grower		Breeder		Horse	0.1
Lamb	0.1	Finisher		Layer		Fish	
Sheep	0.1-0.2	Sow/Boar		Turkey		Rabbits	

Rums	Typical	Pigs	Typical	Poultry	Typical	Others	Typical
Calf	0.1-0.15	Creep	0.5-1	Chick	0.25	Dog	0.5-1
Dairy	0.2-0.5	Weaner	0.5	Broiler	0.5	Cat	0.5
Beef	0.1-0.5	Grower	0.5	Breeder	0.1-0.5	Horse	0.1-0.5
Lamb	0.1	Finisher	0.5	Layer	0.25	Fish	0.25
Sheep	>0.6	Sow/Boar	0.1	Turkey	0.5	Rabbits	

Notes
EU regulation: maximum inclusion in feedstuffs for producing animals: Poultry, Pig, Beef, Dairy: 2mg/kg;

- Direct cobalt requirement is only found in certain bacteria and algae
- Cobalt requirement for ruminants is for ruminal microorganisms. Rumen can synthesise B-vitamins from 6-8 weeks of age
- No NRC recommendations for monogastric species as not required when adequate dietary vitamin B_{12}

Cobalt

8c

| Co | Atomic Wt 58.93 | Atomic No 27 | Micro Mineral |

Adequate Status for Cobalt

Species	Liver	Kidney	Blood	Milk	Serum
Cattle	0.02-0.085	0.071	0.15	0.04-1.1	
Sheep	0.02-0.085	0.071	0.15	0.04-1.1	
Pigs	1.0-2.0	0.4			0.17-0.60
Poultry	0.26-0.85	0.5-1.1			
Units	ppm WW	ppm WW	ppm WW	mg/l DM	ppm WW

<u>Notes</u>
Urine adequate status for Cobalt in Pigs is 16-20 mg/l

Adequate Status for Vitamin B_{12}

Species	Liver	Kidney	Milk	Serum
Cattle	0.25-2.5	0.5-2.6	0.4-0.9	2-4
Sheep	0.3-2.24		1-3.5	
Units	ppm WW	ppm WW	ng/ml DM	µg/l DM

Some indirect tests have been used, (i.e) measuring accumulation of metabolites excreted in urine. The data below is from sheep.

	Co Deficient	Co Adequate
Plasma -methylmalonic acid	5.6 - 6 µg/ ml	1 - 5 µg/ ml
Urine -formininoglutamic acid -methylmalonic acid	0.05 - 1.8 µmol/ ml 30 - 150 µg/ ml	0 - 0.01 µmol/ ml <25 µg/ ml
Serum -formininoglutaminc acid -pyruvate -pyruvate kinase	0.1 - 0.2 µmol/ ml 1 - 2.2 mg % 200 - 5000 mU/ml	0 µmol/ ml 0.6 - 0.9 mg % 40-80 mU/ml

- Levels in the rumen fluid are normally above 20-40 ng cobalt/ml for good Vitamin B_{12} production
- Ruminal fluid Co <0.5ng/ml inhibits vitamin B_{12} synthesis
- Blood vitamin B_{12} concentration correlates well with cobalt status in sheep and goats but not cattle (from assay problem, check with laboratory)

CONTEXT www.contextbookshop.com

Cobalt

Co | Atomic Wt 58.93 | Atomic No 27 | Micro Mineral

Deficiency

General
- Pine (Denmark disease, Coast disease, Enzootic marasmus, Bush sickness, Wasting disease, Nakuritis.)
- Associated with granite and coarse sandy soil areas
- Loss of appetite and retarded growth
- Severe body condition loss
- Abortion
- Muscular weakness with demylination of nerves
- No oestrus/ reduced conception rates
- Rough hair
- Scaliness of the skin
- Increased susceptibility to infection and parasitic infestations
- Fatty degeneration of the liver
- Poor mucous membranes
- Anaemia
- Eventual death
- Deficiency of vitamin B_{12} induces a folacin deficiency. Plasma thyroxine increases probably affecting function of hypothalamus
- Grazing lambs most sensitive to deficiency, then mature sheep, then calves, then mature cattle

Cattle
- Impaired milk production leads to weak calves
- Reduced synthesis and reserves of Vitamin B_{12}
- High concentrate diets will cause the animal to produce less Vitamin B_{12} production and therefore require more cobalt than normal
- Deficiency of Vit B_{12} will affect gluconeogenesis and so glucose supply to tissues
- Limits methionine production and nitrogen retention
- Propionate clearance from blood depressed; high blood propionate reduces VFI

Poultry
- Often offset by common feeds used that usually contain substantial quantities of Vitamin B_{12}

Sheep
- Impairs milk production – weak lambs
- Reduced synthesis and reserves of Vitamin B_{12}
- Worms have a higher requirement for cobalt and therefore will deplete the level in sheep with worms
- 'White liver disease' associated with Co/B_{12} deficiency, possibly coupled with toxic plant metabolite or mycotoxin
- Reduced oestrus, increased still births and neonatal mortalities
- May reduce liver Cu storage and elevate serum Cu levels
- Lambs from deficient ewes slower to start suckling

Toxicity

General
- Damage to heart, kidneys and nerves
- Loss of appetite and body weight
- Anaemia – depresses iron absorption
- Toxic when intake is high, but unusual to be a problem as absorption is usually low
- Hyperchromemia
- Increased liver Co

Pigs
- Toxic diet >100mg/kg LW for 3 days
- Stops eating, gaunt, stiff legs, uncoordinated muscle tremors, hump back
- Myocardial necrosis can occur
- Toxicity aggravated by Fe deficiency
- Toxicity reduced by sulphur amino acids. Cysteine most effective
- Duodenal coccidiosis exacerbates toxicity by increased Co deposition in the liver

Ruminants
- Production of Vitamin B_{12} analogues
- Excessive urination, defaecation and salivation
- Shortness of breath
- Increased haemoglobin, red cell count and packed cell volume
- Sheep less susceptible to toxicity than cattle e.g. young cattle toxic at 1mg/kg LW, sheep 4mg/kg LW

Cobalt

8e

| Co | Atomic Wt 58.93 | Atomic No 27 | Micro Mineral |

Maximum Dietary Tolerable Level (ppm DM)

Cattle	10	Pigs	10
Sheep	10	Rabbits	10
Poultry	10	Horses	10

Interrelationships:
- Cobalt is a component of Vitamin B_{12}
- Deficiency reduces storage of Cu in bovine liver
- Co and Fe mutually inhibit absorption of each other

Antagonists
- Soil pH, molybdenum, liming Possible Mn, Zn and Fe antagonism- can help reduce toxicity
- Zinc or copper can form Vitamin B_{12} analogues
- Co toxicity aggravated by Fe deficiency; compete for absorption sites in the intestine

Synergy
- Co can substitute for zinc in some enzymes e.g carboxypeptidase
- Sulphur amino acids reduce toxicity - poultry

Main Supplements

Source	Element %	Relative Bio-avail.	Comments
Cobalt carbonate	46-55	high	Can convert to oxide (a)
Cobalt chloride	24.7	high	
Cobalt oxide	70	low	(b)
Cobalt glucoheptonate	4	medium	
Cobalt acetate	23		
Cobalt sulphate	21-33	high	Cakes under long storage (c)
Cobalt nitrate	20		

Notes
(a) can convert to cobalt oxide during storage and under exposure to air.
(b) often used in cobalt bullets where slow release solubility is beneficial.
(c) Product needs "Skull & cross bones" and "May cause cancer by inhalation" at levels greater than 0.01% (0.1g/kg) of cobalt sulphate (21mg/kg cobalt from sulphate). Premixes and on farm minerals will often be greater than 0.01% so often use cobalt carbonate.

CONTEXT   www.contextbookshop.com

Cobalt

8f

| Co | Atomic Wt 58.93 | Atomic No 27 | Micro Mineral |

Feed Name	mg/kg DM
Alfalfa meal	0.2
Bakery waste	1.1
Barley grain	0.1
Bean field	0.4
Blood meal	0.1
Brewers grains	0.1
Buckwheat grain	0.1
Buttermilk dehyd.(cattle)	0.01
Casein dehyd. (cattle)	-
Cassava tubers dehy	0.1
Citrus pulp dried	0.2
Copra meal	0.2
Cottonseed whole	-
Cottonseed meal	0.1
Distillers grains - wheat	0.2
Distillers grains maize	0.1
Distillers grains - barley	0.1
Fishmeal (Sth Am)	0.1
Grass bluegrass	-
Grass alfalfa	0.1
Grass bermuda	0.1
Grass clover	0.1
Grass extensive	0.1
Grass kikuyu	0.1
Grass timothy	0.1
Groundnut ext	0.1
Hay alfalfa	0.2
Hay bluegrass	-
Hay Clover	0.2
Hay eragostus	-
Hominy feed	0.1
Linseed meal (mech ext)	0.4
Maize bran	-
Maize germ ext(sol)	0.1
Maize gluten 20	0.02
Maize gluten 60	0.1
Maize grain	0.1
Malt culms	0.1
Milk (cattle-dehyd)	0.01
Milk skimmed	0.1
Millet grain	0.03
Molasses -beet	0.7

Feed Name	mg/kg DM
Molasses - cane	1.00
Oat groats	0.01
Oat middlings	0.1
Oatfeed	-
Oats grain	0.1
Palm kernel exp	0.1
Peas	0.1
Potato dried	0.1
Rape ext (mech)	0.2
Rice bran	0.1
Rice grain	0.04
Rye grain	0.04
Safflower ext. solv.	2.1
Sesame ext mech	0.8
Silage alfalfa	0.3
Silage grass	0.1
Silage maize	0.1
Silage sorghum	0.3
Silage wholecrop	0.1
Sorghum grain	0.2
Soya ext.solv	0.3
Soya flour	-
Soya hipro	0.2
Straw barley	0.1
Straw Oat	0.1
Straw wheat	0.1
Sugar beet pulp (dehyd)	0.1
Sugar beet pulp (mol)	0.2
Sunflower ext	0.2
Triticale grain	-
Wheat (caustic)	0.1
Wheat bran	0.1
Wheat feed	0.1
Wheat germ ext.	0.1
Wheat grain	0.1
Whey low lactose	-
Whey (cattle dehyd)	0.1
Yeast (brewers dehyd)	0.2
Yeast (torula dehyd)	0.03

CONTEXT www.contextbookshop.com

Bioplex® Cobalt from Alltech®

| Co | Atomic Wt 58.93 | Atomic No 27 | Micro Mineral |

Introduction
- Bioplex® Cobalt is an organic trace mineral proteinate for use in livestock feeds.

Concentrations Available

Product	Guaranteed Analysis	Ingredients
Bioplex® Co 2.5%*	Min. 2.5% Cobalt	Cobalt proteinate

*Not all mineral concentrations are available in every country. Contact your local Alltech representative for details.

Physical Characteristics
Appearance
Bioplex® Cobalt is a purple powder with no discernible odour.

Storage
Bioplex® Cobalt should be stored in a closed container in a cool, dry place. Shelf life under these conditions is 36 months.

Packaging
Bioplex® Cobalt is available in 25 kg bags.

Inclusion Rates of Bioplex® Minerals
- Every species requires a different inclusion rate. Also inclusion depends on the motivation for organic supplementation. There are two issues to be examined.
- Organic supplementation can be for performance reasons, here research shows that supplementation with Bioplex® minerals at rates between 20-50% of inorganic minerals provides superior performance effects.
- Supplementation may be driven by the requirement to meet environmental regulations. Bioplex® Cobalt can be used as a complete replacement of inorganic salts at a significantly lower inclusion level.

For inclusion recommendations for your region and other details, contact your local Alltech representative.

Benefits (Conditions responsive to improved Cobalt status)
- Cobalt is required to synthesise vitamin B_{12} in ruminants
- Maintains health of nervous system
- Increases the assimilation of iron and the building of red blood cells
- Cobalt stimulates many enzymes of the body and normalises the performance of body cells

Some Symptoms of Cobalt Deficiency
- Anorexia (loss of appetite)
- Slow growth rate
- Poor production
- Fatigue
- Myelin sheath and nerve damage
- Reproductive disorders

www.alltech.com

Copper

9a

| Cu | Atomic Wt 63.54 | Atomic No 29 | Micro Mineral |

Introduction
- Named from the Greek word kyprios, (Cyprus, where copper deposits were mined by the ancients)
- Chemical symbol is derived from the Latin name for 'copper', 'cuprium'
- A metal that is malleable and ductile
- Natural copper (also called native copper), as a mineral, is relatively rare. In nature, it easily combines with a number of elements and ions to form a wide variety of copper minerals and ores. Deposits large enough to mine include azurite, malachite, tennantite, chalcopyrite and bornite
- Used in electric cables and wires, switches, plumbing, heating, roofing and building construction, chemical and pharmaceutical machinery, alloys (brass, bronze), alloy castings, electroplated protective coatings and undercoats for nickel, chromium, zinc, etc., and cooking utensils
- Used in many agricultural products, e.g. plant and animal fungicides, foot baths
- Concentration in earths crust about 50ppm
- An essential metal element, for animals, for blood formation and through enzyme complexes in many metabolic processes
- 10th most abundant mineral found in the body
- Animal's body contains approx. 2-3mg/kg copper
- Within animals large amounts of available copper migrate to the liver, smaller amounts are found in blood, bone and soft tissues – the amounts are strongly related to dietary concentration
- Copper can also be a toxicant

Key Natural Sources
- Crop and forage concentration varies geographically
- Plant uptake is influenced by species, stage of maturity, yield, crop management, climate, geography, drainage, pH, soil type, antagonists (e.g. Mo,S,Fe), organic matter (muck)
- Free draining sandy soils tend to have low copper contents and clay-type soils have the highest contents. e.g Sandstones 10-40ppm; shales 30-150ppm, marine black shales 20-300ppm
- Cu concentration tends to decline as plants mature Poor drainage, acid pH, high clay, low antagonists, low organic matter encourage uptake.
Levels range from 1-50ppm DM
- Copper can accumulate in plants and animals when it is found in soils
- Cu level and availability of grains tends to be higher than leaves and stems
- Availability of Cu in feeds varies (e.g.) hay > fresh grass > silage
- Copper water pipes can contribute to Cu intake particularly acid, soft water areas
- Milk is low in copper; so animals on a predominant natural milk diet can develop anaemia

Function
- Blood haemoglobin and myoglobin synthesis, with iron. (oxygen carrying components in the blood)
- Component of or essential in the activity of many enzymes. e.g. cytochrome oxidase, catalase, tyrosinase, ascorbic acid oxidase etc
- Plays a key function in oxido-reductase enzymes e.g Superoxide dismutase, protecting cells from oxygen metabolites toxic effects
- Cu-enzyme catalyses formation and inactivation of some hormones
- Involved in the development and maintenance of vascular and skeletal systems e.g. blood vessels, tendons and bones. (Necessary for collagen, elastin, bone & connective tissues)
- Involved in oxygen metabolism
- Involved in the body's 'metal' management through its presence in ceruloplasmin
- Aids the absorption of iron from the intestine (part of ferroxidase enzyme) and releases it from storage in the liver and reticuloendothelial system
- Needed for structure & function of the central nervous system
- A key component in the pigmentation of hair and wool (turacin, a feather pigment)
- Necessary for enzyme in production of melanin pigment from L-tyrosine
- A key component of keratinisation of hair, wool and feathers
- Component of copper containing proteins

CONTEXT   www.contextbookshop.com

Copper

| Cu | Atomic Wt 63.54 | Atomic No 29 | Micro Mineral |

Benefits
- Bone formation
- Proper cardiac function
- Immune system
- Cell respiration
- Important in pigments
- Reproduction/fertility
- Increased growth rate and feed efficiency in young pigs and to a lesser extent in broilers. (possible anti-microbial action in gut)

Absorption
- Absorption by active transport that is saturable and simple diffusion that is unsaturable
- Generally poor and in all segments of the gastrointestinal tract, majority tends to be from upper small intestine
- Varies by species, stage of development, chemical form and interactions with other dietary factors
- Absorption greater in young than mature animals. Adult ruminants very low absorption, 1-5%; pre-ruminants and monogastrics 47% reported
- Cu bioavailability affected by genetics of ruminants
- Homeostasis affected by controlling rate of absorption that is regulated by intestinal mucosa in relation to need
- Metallothionein (Cu- binding protein) binds Cu in epithelial cells of intestine, prevents serosal transfer and so may play a role in regulation
- May help protect against overdose

Influence of total dietary sulphur and molybdenum on absorption of copper in sheep.
(Suttle and McLauchlan, 1976 as shown in ARC, 1980)

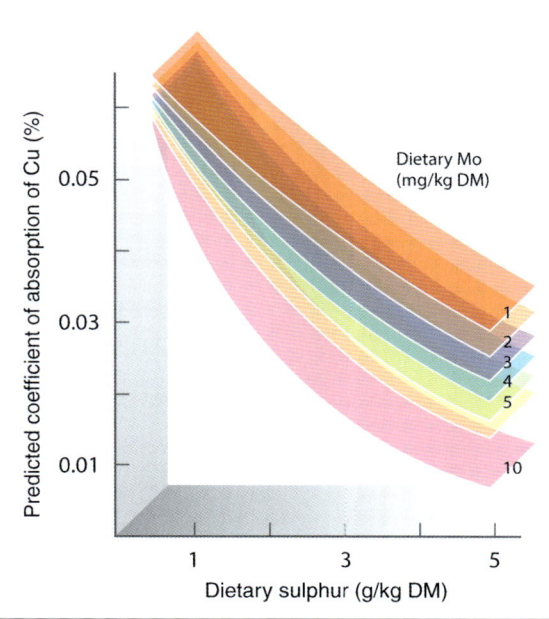

Metabolism
- Absorbed copper enters the blood with 80-90% becoming bound to globulin –copper complex called ceruloplasmin. The remainder is loosely bound to albumin and transported in the blood to the tissues
- Ceruloplasmin is synthesised in liver and in white blood cells.
- It is carrier of Cu from liver to target organs. Cu is quickly deposited in liver
- Liver is central organ of metabolism
- Levels reflect intake, interactions and Cu status
- Stored in liver as a metallothienen complex. Liver stores can be used in times of dietary inadequacy

www.contextbookshop.com

Copper

9c

| Cu | Atomic Wt 63.54 | Atomic No 29 | Micro Mineral |

Excretion
- High proportion of ingested Cu lost via faeces as unabsorbed copper
- Active excretion via the bile. Intermediate levels excreted through urine, milk, intestines
- Small amount via sweat
- Maximum levels of copper in feeds are restricted because of the resultant slurries being potentially toxic to other livestock.
- Phytate can increase excretion

Requirement /Allowances (mg/kg DM)

Rums	NRC	Pigs	NRC	Poultry	NRC	Others	NRC
Calf	10	Creep	6	Chick	5	Dog	2.9
Dairy	9-18	Weaner	6	Broiler	8	Cat	5
Beef	10	Grower	4	Breeder	8	Horse	10
Lamb		Finisher	3.5	Layer	6	Fish	3.5
Sheep	7-11	Sow	5	Turkey	6-8	Rabbits	2.5-8

Rums	Typical	Pigs	Typical	Poultry	Typical	Others	Typical
Calf (MR)	10	Creep	4-6	Chick	4-10	Dog	4-7
Dairy*	10-20	Weaner#	150-250	Broiler	8	Cat	5-15
Beef*	10	Grower	175	Breeder	8	Horse	10
Lamb	1-3	Finisher	100	Layer	6	Goats	12
Sheep	5-10	Sow/Boar	10	Turkey	8		

Notes
Young pigs often fed 100-250 ppm to stimulate growth
* Dietary Mo and S levels affect requirement for Cu
<u>EU regulation</u>: maximum inclusion in feedstuffs for producing animals:
Poultry- 15mg/kg, Pig – 25mg/kg, Beef, Dairy -35mg/kg

Adequate Status

Species	Liver	Kidney	Serum	Milk	Hair
Cattle	25-100		0.8-1.5	0.05-0.6	6.7-32
Sheep	25-100	4-5.5	0.7-2	0.2-1.5	2.8-10
Goats	25-100	3-6	0.4-0.8	0.1-0.9	
Pigs	5.25	7-10	1.3-3	1.18-1.6	8-15
Layers	3-6	3-4.8	0.2-0.45		
Broilers	3-15	3-4.8	0.08-0.18		
Horses	4-7.5	7.3-9.3	0.85-2.0		5-13
Rabbits	8-50	4-6	2-2.4		
Cats	37-45	2.2-2.7	0.6-1.4		
Dogs	30-100	5-15	0.2-0.8dm		
Units	ppm WW	ppm WW	ppm WW	mg/l DM	ppm DM

- Cattle: serum levels a quick indicator of advanced deficiency status but not reliable after supplementation or marginal deficiency
- Liver Cu more informative but may be affected by other dietary interactions
- Milk levels vary with species, stage of lactation and Cu nutrition of the animal

CONTEXT www.contextbookshop.com

Copper

9d

| Cu | Atomic Wt 63.54 | Atomic No 29 | Micro Mineral |

Deficiency

General
- Anaemia
- Abnormal appetite
- Reduced growth rate
- Keratinisation failure in hair, fur, wool
- Abnormal skeletal development
- Reduced protein synthesis
- Nerve disorders (e.g. neonatal ataxia)
- Lameness, swelling of joints
- Diarrhoea
- Weak, broken bones
- Cardiovascular disorders
- Reproductive failure
- Severe deficiency: loss of integrity of elastic and connective tissue

Cattle
- Hair depigmentation especially around the eyes
- Affects the nervous system
- Poor fertility – delayed or depressed oestrus, abortion, poor semen quality
- Retained placenta
- Decreased milk production
- Impaired immune response: poor vaccination response, severe parasitism, failure to respond to treatment
- Deficiency is simple or induced (high Mo & S)

Pigs
- Bowing legs, spontaneous fractures
- Young pigs- ataxia

Poultry
- Chicks: easily broken bones – lack of collagen cross-linking
- Layers: prolonged deficiency reduces egg production and hatchability, abnormal eggs

Sheep
- Depigmentation of the wool, coarse and straight = steely wool
- Lambs: swayback (neonatal ataxia) from low copper status ewe: 'common acute' occurs in newborns or 'delayed' weeks or months after birth. If severe, lambs are born paralysed and die immediately

Dogs
- Rare if on meat diet as Cu efficiently utilised
- Achromotrichia, depigmentation of hair has been observed

Horses
- Low Cu, high Zn diets increase development of osteochondrosis and skeletal abnormalities in youngstock

Toxicity

General
- Tolerance species dependent and breed differences
- Ruminants most sensitive to toxicity, vomiting, salivation, abdominal pain, convulsions, paralysis, collapse, death
- Can impair appetite and digestion, decreased growth
- Emaciation, muscular dystrophy
- Impaired reproduction
- anaemia
- Liver damage, jaundice, bloody urine, necrosis, death

Ruminants
- Calves less tolerant than adults
- Veal calf performance is impaired due to clinical diarrhoea, pneumonia, septicemia

Pigs
- Occasional from 250ppm Cu as result of lack of Zn and Fe

Poultry
- High levels increase need for sulphur amino acids
- Proventriculitis in young broilers can act as a pro oxidant – destroys fat and fat-soluble components of diet

Sheep
- Specific sheep breeds are very susceptible to copper toxicity, death may result

Goats
- 3 times more resistant to toxicity than sheep. Low Mo intake accentuates susceptibility to Cu toxicity

Horses
- Horses tend to be more tolerant than other species

Copper

9e

| Cu | Atomic Wt 63.54 | Atomic No 29 | Micro Mineral |

Maximum Dietary Tolerable Level (ppm DM)

Cattle	100	Pigs	200
Sheep	25	Rabbits	250
Pigs	250	Horses	800
Poultry	300		

Interrelationships
- Cu required for proper Fe metabolism.
- Cu and Se deficiency frequently occur concurrently
- Stress can release large quantities Cu from the liver into the blood

Antagonists
- Excess Cu interacts with Ca, Zn, Fe, Mo, S
- Absorption reduced by: high dietary phytates, fructose, nitrate/protein, plant oestrogens, high Ca, S, Fe, Zn, Cd, Ag, Mo
- Intestinal binding agents (e.g.. bile, fibre) inhibit absorption
- Copper utilisation affected by: Mo, S, Ca, Zn, Co, Cd, Fe, Hg, Ag, Se, plant oestrogens
- High dietary Zn reduces liver stores of Fe and Cu
- Utilisation decreased by ascorbic acid, high dietary nitrate/ protein
- In ruminants: high S feeds react with molybdate forming thiomolybdates in rumen. These complex with Cu into non-utilisable form, $CuMoSO_4$
- Cu-Mo-S interaction is complex and occurs primarily in digestive tract but also sites of metabolism. Ruminants more susceptible than non-ruminants:

1) Mo, and especially Mo in presence of S (formation of insoluble thiomolybdates), reduces Cu absorption and deposition to organs and synthesis of ceruloplasmin Urinary Cu increases and Cu excretion with bile decreases
2) Increasing dietary Cu reduces Mo deposition in liver
3) Increasing dietary S, increases urinary Mo and Mo tissue deposition decreases
- Mo >1ppm dm depresses Cu absorption, >5ppm inhibits Cu storage and molybdenosis result
- Dietary Cu: Mo ratio for normal performance 5:1 to 10:1. High Mo increase requirement for Cu

Synergy
- Selenium supplementation tends to reduce serum Cu (ceruloplasmin) levels – this seems to be a sparing effect
- Cu aids iron absorption
- Chelating agents (amino acids, citrate) enhance Cu absorption

Main Supplements
- Provision via feeds, mineral supplements, drenching, slow release bolus

Copper

9f

| Cu | Atomic Wt 63.54 | Atomic No 29 | Micro Mineral |

Main Supplements

Source	Element %	Relative Bio-avail.	Comments
Copper hydroxide			
Copper lysine	10	high	
Copper nitrate	33.9	med	
Copper orthophosphate			
Copper sulphate	25	high	Water soluble High hygroscopicity
Copper acetate		high	Water soluble
Cupric carbonate	53	high	Water insoluble Low hygroscopicity
Cupric chloride	37.2	high	
Cupric oxide	75-80	low	Water insoluble Low hygroscopicity Often fed as needles or slow release in rumen-reticulum
Organic copper	8.5-15	high	

Copper

9g

| Cu | Atomic Wt 63.54 | Atomic No 29 | Micro Mineral |

Feed Name	mg/kg DM	Feed Name	mg/kg DM
Alfalfa meal	10.8	Molasses - cane	64.0
Bakery waste	5.0	Oat groats	6.8
Barley grain	7.0	Oat middlings	4.4
Bean field	11.1	Oatfeed	2.2
Blood meal	11.0	Oats grain	4.8
Brewers grains	22.2	Palm kernel exp	27.5
Buckwheat grain	11.0	Peas	7.7
Buttermilk dehyd.(cattle)	1.0	Potato dried	7.7
Casein dehyd. (cattle)	4.0	Rape ext (mech)	6.7
Cassava tubers dehy	4.5	Rice bran	14.5
Citrus pulp dried	5.7	Rice grain	3.3
Copra meal	15.0	Rye grain	8.0
Cottonseed Whole	54.9	Safflower ext. solv.	11.0
Cottonseed meal	24.0	Sesame ext mech	35.0
Distillers grains - wheat	21.0	Silage alfalfa	9.0
Distillers grains maize	40.0	Silage grass	7.0
Distillers grains- barley	49.5	Silage maize	5.0
Fishmeal (Sth Am)	9.8	Silage sorghum	35.0
Grass bluegrass	14.0	Silage wholecrop	8.0
Grass alfalfa	10.0	Sorghum grain	11.0
Grass bermuda	6.0	Soya ext. solv	19.1
Grass clover	9.0	Soya flour	-
Grass extensive	7.0	Soya hipro	18.2
Grass kikuyu	7.0	Straw barley	5.0
Grass timothy	11.0	Straw oat	10.0
Groundnut ext	15.0	Straw wheat	3.4
Hay alfalfa	11.0	Sugar beet pulp (dehyd)	14.0
Hay bluegrass	10.0	Sugar beet pulp (mol)	12.4
Hay clover	11.0	Sunflower Ext	37.3
Hay eragostus	5.0	Triticale grain	11.5
Hominy feed	15.5	Wheat (caustic)	5.8
Linseed meal (mech ext)	24.0	Wheat bran	12.4
Maize bran	11.2	Wheat feed	9.0
Maize germ ext(sol)	4.4	Wheat germ ext.	11.0
Maize gluten 20	2.3	Wheat grain	5.8
Maize gluten 60	7.8	Whey low lactose	8.0
Maize grain	24.0	Whey(cattle dehyd)	50.0
Malt culms	14.4	Yeast (brewers dehyd)	35.0
Milk (cattle-dehyd)	1.0	Yeast (torula dehyd)	14.0
Milk skimmed	1.05		
Millet grain	28.0		
Molasses -beet	16.5		

CONTEXT www.contextbookshop.com

Bioplex® Copper from Alltech®

| Cu | Atomic Wt 63.55 | Atomic No 29 | Micro Mineral |

Introduction
- Bioplex® Copper is an organic trace mineral proteinate for use in livestock feeds.

Concentrations Available

Product	Guaranteed Analysis	Ingredients
Bioplex® Cu 10%*	Minimum 10% Copper	Copper proteinate

*Not all mineral concentrations are available in every country. Contact your local Alltech representative for details.

Physical Characteristics
Appearance
Bioplex® Copper is a blue-green coloured, free flowing powder with no discernible odour.

Storage
Bioplex® Copper should be stored in a closed container in a cool, dark area. Shelf life under these conditions is 36 months.

Packaging
Bioplex® Copper is available in 25 kg bags.

Inclusion Rates of Bioplex® Minerals
- Every species requires a different inclusion rate. Also inclusion depends on the motivation for organic supplementation. There are two issues to be examined.
- Organic supplementation can be for performance reasons, here research shows that supplementation with Bioplex® minerals at rates between 20-50% of inorganic minerals provides superior performance effects.
- Supplementation may be driven by the requirement to meet environmental regulations. Bioplex® Copper can be used as a complete replacement of inorganic salts at a significantly lower inclusion level.

For inclusion recommendations for your region and other details, contact your local Alltech representative.

Benefits (Conditions responsive to improved Copper status)
- Copper increases iron assimilation
- Copper and iron work together in the formation of haemoglobin and red blood cells
- Influences the immune system
- Copper improves vitamin C oxidation and is integral in the formation of RNA
- Involved in proper vascular and skeletal growth and function

Some Symptoms of Copper Deficiency
- Retarded growth
- Impaired muscle coordination
- Anaemia
- Bone and cardiovascular disorders
- Reproductive failure

Alltech® ...naturally

www.alltech.com

Fluorine

10a

| F | Atomic Wt 19 | Atomic No 9 | Minor Mineral |

Introduction
- Fluorine, a highly reactive, nonmetallic, pale yellow gas of halogen chemical group
- Fluoride occurs naturally in the earth's crust where it can be found in rocks, coal and clay
- Fluorides are released into the air in wind-blown soil
- Hydrogen fluorides can be released into air through combustion processes in the industry
- In the environment fluorine cannot be destroyed; it can only change form
- Discovered in 1886 by Joseph Henri Moissan
- Name 'Fluorine' from the Latin word 'fluo' (flow)
- Atmospheric contaminants from industry (phosphate ore processors, aluminum, steel and copper smelters, some chemical manufacturers, brick or ceramic product factories)
- In animals it is found particularly in the bones and teeth but also has beneficial effects in the blood
- A very toxic element

Key Natural Sources
- Found naturally in soil, water, plant and animal tissues
- A key component of rock salts
- Defluorinated, feed grade phosphates must contain a ratio of less than 1:100 fluorine to phosphate
- Some rock phosphates contain up to 4% F
- The uptake by plants depends upon the type of plant, soil and the amount and type of fluorine found in the soil. (high levels in soil surrounding volcanoes)
- Most plants have a limited absorption capacity from soil
- Liming (increasing soil pH) reduces plant F uptake
- Fluorine contamination of fertilisers can affect plant fluorine levels
- Large amounts in shellfish and tea. Fishmeal levels tend to be over 200mg/kg (salt water fish higher than fresh water)
- Historically blood and bone meal were also good sources (100 mg/kg) but their usage in some markets are now restricted
- Water F levels vary by geographical location. In many countries F is added to the water for humans and therefore also supplemented to livestock
- Contaminated plant materials and phosphate supplements can influence livestock levels

Function
- Forms up to 0.05% of teeth and bones and found especially on the surface
- It has an effect on calcium metabolism and potentially in many other areas of the body's metabolism including energy metabolism
- Known to reduce iron, magnesium, calcium and chlorine and even lipid absorption in the small intestines
- Potentially a cofactor in many enzyme reactions including amylase where calcium and magnesium may also be involved but large levels required

Benefits
- Small amounts beneficial for dental health
- May help prevent calcification of organs and muscoskeletal structures
- May help prevent bone decalcification in older animals (osteoperosis)

Absorption
- 90% of ingested fluorine is absorbed by passive absorption from the upper intestines
- Domestic animals and humans, F absorbed from stomach and rumen in ruminants
- High gastric acidity increases F absorption
- Solubility of ingested material plays important role in absorption
- Absorption is rapid (half-time about 30minutes)

CONTEXT www.contextbookshop.com

Fluorine

| F | Atomic Wt 19 | Atomic No 9 | Minor Mineral |

Metabolism
- F in blood instantaneously reacts with Ca to form calcium fluoride
- Plasma levels maintained within narrow limits by regulatory mechanisms in kidney activity and increased bone deposition
- F level is regulated by the kidneys with up to 40% of absorbed material excreted within 5 hours
- Fluoride does not accumulate in soft tissues
- F accumulates in rib and mandible bone as a function of age
- When bone levels increase to excess, metabolic breakdown results with reduction in appetite and eventually death
- F can cross the placental barrier

Excretion
- Excretion mainly in the urine (90%) with a little in the faeces and sweat
- Dietary levels can affect F in milk, eggs (especially yolk)
- High F levels, urinary F and skeletal tissue deposition increases

Requirement
- Generally no need for supplementation, normal feedstuffs contain sufficient levels
- Dietary F sources are additive in animals and so collective total should be assessed

Adequate Status

Species	Liver	Kidney	Blood	Milk	Serum
Cattle (a)			0.1-0.3	0.01- 0.06	
Pigs	0.3	1.2			6.5-61
Poultry	0.2 - 4.8	2.9 - 3.8			
Units	ppm DM	ppm DM	ppm WW	ppb WW	% DM

Notes
(a) Urine level typically 1.0-5.0 ppm WW

- No single criterion should be relied on for diagnosing and evaluating F toxicosis
- Age and bone content give indication of potential damage
- Plasma levels relate to current ingestion. Urine level related to dietary intake

Deficiency

General
- Not common in farm livestock under natural conditions
- Symptoms may include:
  - Growth rates and number of offspring could be reduced
  - Osteoporosis (bone decalcification), infertility and anaemia
  - Dental caries increases

Fluorine

10c

| F | Atomic Wt 19 | Atomic No 9 | Minor Mineral |

Toxicity

General
- A cumulative poison and therefore fluorosis may not be seen for some time
- Once bone tissue is saturated, continued intakes deposited in soft tissues, metabolic disturbance and death result

Clinical Signs
- Longer-lived animals are more susceptible to chronic fluorosis
- Bone and teeth thickening
- Teeth become discoloured, pitted and deformed and suffer from irregular wear
- Secondary and tertiary hyper eparathyroidism
- Joints thicken and the long bones of animals may shorten
- Lameness, stiffness, treading of feet, curled and abnormal hoofs
- Roughened coat
- Delayed maturity
- Poor feed utilisation
- Appetite and growth is reduced
- Possible reduced immune response
- Delayed oestrus

Ruminants
- More susceptible than non-ruminants
- Ruminal decrease in volatile fatty acids and protozoal population
- Reduced milk production

Pigs
- Less susceptible than ruminants and horses

Poultry
- Least susceptible
- Egg production and intake depressed diets >950ppm F from rock phosphate
- Hatchability may be reduced as F transfers to eggs
- Acute toxicity reduces serum Na, Ca, P and raises serum K

Horses
- Less susceptible than cattle or sheep
- Appear unthrifty and apparent pain when standing

Maximum Recommended Levels: (as sodium fluoride or fluorides)

	Diet ppm dry matter	Drinking water (mg/l)
Young dairy cattle	40	2.5-4.0
Slaughter cattle	100	12-15
Mature dairy cattle	40	3-6
Mature beef cattle	50	4-8
Ewes	60	5-8
Finishing lambs	100-150	12-15
Horses	40	4-8
Pigs	150	5-8
Chicken	200	2
Turkeys	150	2
Rabbit	40	

Interrelationships
- F interacts with Al, Ca, P, I and possibly many other elements
- F may interfere with Mg, Mn, Fe, Mo
- Cu and Zn metabolism

Antagonists
- Absorption reduced by Al, Ca, Mg, B and increased by fats
- Low Ca intake results in increased F uptake
- Protein absorption decreases with increasing dietary F
- Vitamin B_{12} synthesis and folic acid activity are compromised
- F affects ascorbic acid levels in young chicks
- Toxic levels interfere with Ca metabolism

Fluorine

| F | Atomic Wt 19 | Atomic No 9 | Minor Mineral |

Synergy
- Phosphates and fluorine may complex in the bones but it does not appear to affect its F deposition
- Inadequate dietary carbohydrate intake enhances F absorption. (ruminants)
- High F and Mg synergistically reduce growth rate and bone ash (poultry)
- Al (as sulphate, chloride, lactate, hydroxide) reduces F toxicity and accumulation in bone. - sheep

Typical Composition of Phosphate Compounds

	Fluorine %	Phos %	P:F ratio	F (a) (ppm)
Phosphates manufactured from defluorinated wet-process phosphoric acid				
Monocalcium/dicalcium phosphate	0.16	21.0	131	19
Dicalcium/monocalcium phosphate	0.14	18.5	132	19
Defluorinated phosphate	0.16	18.0	113	22
Monoammonium phosphate	0.18	24.0	133	19
Diammonium phosphate	0.16	20.0	125	20
Ammonium polyphosphate solution	0.12	14.5	121	20
Defluorinated wet-process phosphoric acid	0.18	23.7	132	19
Phosphates manufactured from furnace phosphoric acid				
Monocalcium phosphate	0.03	23.0	767	3
Dicalcium phosphate	0.05	18.5	370	7
Tricalcium phosphate	0.05	19.5	390	6
Monosodium phosphate	0.03	25.5	850	3
Disodium phosphate	0.03	21.5	717	4
Sodium tripolyphosphate	0.03	25.0	833	3
Ammonium polyphosphate solution	0.02	16.0	800	3
Furnace phosphoric acid	0.03	23.7	790	3
High-fluoride phosphates				
Soft rock phosphate	1.2	9.0	7.5	334
Ground low-fluorine rock phosphate	0.45	14.0	31	81
Triple superphosphate	2.0	21.0	10.5	238
Diammonium phosphate (fertiliser grade)	2.0	20.0	10	250
Wet-process phosphoric acid	2.5	23.7	9.5	264
Ground rock phosphate	3.7	13.0	3.5	710

Notes
(a) Contribution to provide 0.25% P

Main Supplements
- No reports of responses to supplementation as levels found in natural feedstuffs meet requirements
- Water sources for agriculture should be checked for F content whether consumed directly or used on forage crops
- Primary source of dietary F are from inorganic phosphates used in mineral supplements
- Most commercial feed-grade phosphates have been chemically processed to remove Flourine
- Relative availability of Flourine, Sodium fluoride 100%, defluorinated phosphate 65%, soft rock phosphate 50-53%, Dicalcium phosphate 18-20%

Fluorine

10e

| F | Atomic Wt 19 | Atomic No 9 | Minor Mineral |

Feed Name	mg/kg DM	Feed Name	mg/kg DM
Alfalfa meal	-	Molasses - cane	-
Bakery waste	-	Oat groats	-
Barley grain	3.2	Oat middlings	-
Bean field	1.1	Oatfeed	-
Blood meal	13.3	Oats grain	1.4
Brewers grains	4.9	Palm kernel exp	3.4
Buckwheat grain	-	Peas	1.2
Buttermilk dehyd.(cattle)	1.1	Potato dried	2.7
Casein dehyd. (cattle)	0.2	Rape ext (mech)	15.0
Cassava tubers dehy	5.1	Rice bran	-
Citrus pulp dried	-	Rice grain	0.2
Copra meal	6.7	Rye grain	1.5
Cottonseed Whole	-	Safflower ext. solv.	-
Cottonseed meal	-	Sesame ext mech	9.8
Distillers grains - wheat	-	Silage alfalfa	-
Distillers grains maize	-	Silage grass	-
Distillers grains- barley	-	Silage maize	-
Fishmeal (Sth Am)	200.0	Silage sorghum	-
Grass bluegrass	-	Silage wholecrop	-
Grass alfalfa	-	Sorghum grain	0.8
Grass bermuda	-	Soya ext.solv	7.8
Grass clover	-	Soya flour	-
Grass extensive	-	Soya hipro	-
Grass kikuyu	-	Straw barley	-
Grass timothy	-	Straw oat	-
Groundnut ext	-	Straw wheat	4.5
Hay alfalfa	18.9	Sugar beet pulp (dehyd)	6.7
Hay bluegrass	-	Sugar beet pulp (mol)	-
Hay clover	-	Sunflower ext	-
Hay eragostus	-	Triticale grain	-
Hominy feed	-	Wheat (caustic)	2.3
Linseed meal (mech ext)	4.1	Wheat bran	4.0
Maize bran	-	Wheat feed	0.3
Maize germ ext(sol)	1.0	Wheat germ ext.	-
Maize gluten 20	-	Wheat grain	2.3
Maize gluten 60	4.4	Whey low lactose	-
Maize grain	-	Whey (cattle dehyd)	0.4
Malt culms	-	Yeast (brewers dehyd)	2.6
Milk (cattle-dehyd)	8.3	Yeast (torula dehyd)	7.5
Milk skimmed	0.6		
Millet grain	-		
Molasses -beet	2.0		

CONTEXT www.contextbookshop.com

Iodine

11a

| I | Atomic Wt 126.9 | Atomic No 53 | Micro Mineral |

Introduction
- A shiny blue-black solid element of the halogen family
- On heating it goes directly to a gas without passing through a liquid phase (sublimes). Iodine gas is toxic
- Discovered by French scientist Bernard Courtois in 1811 when he treated seaweed ash with sulphuric acid
- Seaweed was a major source of iodine before 1959
- Named 'Iodine' comes from the Greek word *'iodes'* which means violet in reference to its colour
- Primarily retrieved from underground brines (water with many dissolved salts and ions) that are associated with natural gas and oil deposits. Also retrieved as a by-product with nitrate deposits in caliche deposits
- Relatively rare in earth's crust containing about 0.0004% by weight
- Chile is the leading iodine producing nation, followed by Japan
- Used in a number of chemical and biological applications. E.g. photography, as a disinfectant, as a catalyst, used to make inks and colourants
- Potassium iodide is included in table salt as a simple way to assure adequate iodine in the human diet
- Used as a supplement in animal feeds
- Iodine deficiency is one of the most prevalent diseases in the world
- It is found throughout the body - about 0.4 mg/kg body weight. 70-80% of the body's reserves are found in the thyroid gland
- The thyroid gland can only produce thyroid hormones if sufficient iodine is available

Key Natural Sources
- Found in most food at trace levels (<0.5mg/kg DM) although marine sources are often the highest e.g. seaweed meal and fishmeal are rich sources
- Seawater contains approximately 50 μg/l iodine, fresh water 5μg/l, soil 300μg/kg
- Soil content is very variable
- Plant uptake dependant on species, soil type and content, fertiliser applications and climatic conditions
- Concentration in common feedstuffs in highly variable
- Coastal regions tend to have soil with a higher Iodine content
- Most pastures tend to be marginal or deficient
- N fertilisation has reduced forage Iodine
- Feed iodine is not stable during processing and storage with over 50% at times subliming
- Brassica species contain goitrogenic substances that affect absorption

Function
- An essential component of thyroid hormones, thyroxine and triiodothyronine that are intimately involved in the control of metabolic activity
- Production of various body enzymes, e.g. in respiratory system, oxidation, phosphorylation and protein synthesis

Thyroid hormones have active role in:
- Thermoregulation
- Growth, development
- Intermediary metabolism
- Reproduction
- Circulation
- Muscle function
- Control rate of oxidation of all cells

Thyroid hormones also:
- Influence physical and mental development
- Influence neuromuscular function
- Affect other endocrine glands, e.g. gonads
- Have an effect on hair, fur, feathers
- Influence metabolism of proteins, carbohydrates and lipids, some minerals and water
- Conversion of carotene to Vitamin A appears to be regulated by thyroid hormones

Benefits
- Required for growth by some rumen microflora
- Helps regulate basal metabolic rate

CONTEXT   www.contextbookshop.com

Iodine

| I | Atomic Wt 126.9 | Atomic No 53 | Micro Mineral |

Absorption
- Very efficient (nearly 100%) of organic and inorganic iodine is absorbed. Usually reduced to iodides
- Absorbed through entire gastro intestinal tract with the majority in the small intestine
- Ruminants: 70-80% of intake is absorbed directly from the rumen. 10% from the abomasum
- Iodide and I readily absorbed in the lungs
- I compounds can be absorbed through the skin
- Reabsorption can occur from saliva, GI fluids, I hormone breakdown etc.
- Goitrogens affect absorption. (see interactions below)

Metabolism
- Iodine enters the blood where it enters the iodide pool. Found in all tissues
- Most is used by the thyroid gland. Thyroglobulin is the storage form and is about 90% of total thyroid I
- In thyroid gland, elemental iodides are released and combine with the amino acid tyrosine to form mono and di-iosotyrosines. These are eventually converted to hormone thyroxine (tri and tetra-iodotyronines), which is released into the blood. This is all from stimulation by pituitary thyroid stimulating hormone (TSH)
- Transport to target tissues (all cells in body) is by thyroid binding protein
- Some deiodination of these hormones occurs in target tissues and also in the kidneys which leads to excretion via urine
- Deiodination in the cell is catalysed by a Se-containing enzyme
- I recycling occurs in ruminant from blood plasma to and from: uterus and conceptus, mammary gland, kidney, liver, lower gi tract, abomasums, forestomachs, saliva, thyroid
- During lactation, thyroxine production increases between 2 and 3 fold in high producing cows
- Iodine freely passes across the placenta and accumulates in foetus tissues and blood
- Low I levels lead to mobilisation of nutrients from stores that leads to weight loss

Excretion
- Up to 70% is excreted as free I in urine and continues even in deficiency
- Unused iodide or sequestered I is excreted via faeces
- Some loss via sweat and especially in tropical areas
- Milk I varies with stage of lactation, intake, body fluid status and environ temperature
- Colostrum I is about 4-5 times greater than milk level

Iodine

| I | Atomic Wt 126.9 | Atomic No 53 | Micro Mineral |

Requirement /Allowances (mg/kg DM)

Rums	NRC	Pigs	NRC	Poultry	NRC	Others	NRC
Calf	(a)	Creep	0.14	Chick	0.35	Dog	
Dairy	0.34-0.88	Weaner	0.14	Broiler	0.35	Cat	0.35
Beef	0.5	Grower	0.14	Breeder	0.44	Horse	0.1
Lamb	0.3	Finisher	0.14	Layer	0.35	Fish	0.6-1.1
Sheep	0.1-0.8	Sow/Boar	0.14	Turkey	0.4	Rabbits	0.2

Rums	Typical	Pigs	Typical	Poultry	Typical	Others	Typical
Calf	(b)	Creep	0.2-0.4	Chick	1.1-1.5	Dog	1-2
Dairy	0.8-2	Weaner	0.4-1	Broiler	1.5-2	Cat	1
Beef	0.5-0.8	Grower	0.4-1	Breeder	1-2	Horse	0.1-0.2
Lamb	0.25	Finisher	0.4-1	Layer	0.44-2	Fish	1
Sheep	0.2-0.5	Sow/Boar	0.2-0.6	Turkey	2-2.5	Rabbits	0.5-1.2

Notes
(a) For milk replacer 0.25 (b) For milk replacer 0.5
Maximum Iodine permitted in feeds 40mg/kg (feed corrected to 12% moisture)
Requirements influenced by climate and environment. Secretion of thyroid hormones in ruminants is inversely related to environmental temperature e.g. higher in winter than in summer. If diet contains 25% strongly goitrogenic feed on a dry basis, iodine should be at least doubled e.g cattle and sheep to 2mg/kg DM

Adequate Status

Species	Liver	Serum-total	Serum T4	Milk	Urine
Cattle	0.094-2.0	10-40	2.5-6.0	20-300	10-25
Sheep		3-12.3	5.5-6.6	80-400	
Goats		10-40	6-10	25-400	
Pigs		9.3-20	3-7.4	38-281	
Poultry		2-10	0.7-3.0		
Horses		2-10	1-2.5		
Rabbits		5-10	2.1-6.2		
Dogs		5-20	1-4		
Units	ppm WW	µg/100ml DM	µg/100ml DM	ppb DM	µg/100ml DM

- Ratio of T_3 to T_4 is also important. Total serum I correlates better with dietary intake than protein bound iodine
- Liver levels do not correlate with dietary intake
- Levels in milk and eggs readily reflect dietary sources
- Typical egg I : 4-12 µg/100g dm with most found in yolk
- Milk levels may be influenced by iodophor antiseptics used in udder washes, milk machines, storage and transportation tanks etc
- Radio immuno assays for thyroid hormones and thyrotropin provide alternative method of assessing thyroid and iodine status
- Blood TSH level is a good indicator of I status

Synergy
- Retinoic acid and thyroid hormone control thought to have some synergy

Iodine

| I | Atomic Wt 126.9 | Atomic No 53 | Micro Mineral |

Deficiency

General
- Thyroid gland becomes over active and enlarges, due to the growth of connective tissue, to give the Goitre (Goiter) appearance. The actual gland becomes smaller with potential atrophy
- Reduced growth rate
- Dry skin, rough, harsh, brittle hair
- Weak or dead, hairless newborns
- Reduced metabolic rate
- Growing animals – shortening of leg bones

Ruminants
- Affects the rumen microflora, sheep have poor wool
- Foot rot (cattle - interdigital phlegmon)
- Retained placenta
- Abnormal oestrus (irregular or suppressed)
- Resorption of foetus (early embryonic death)
- Abortions
- Stillbirths
- Blind, hairless, weak or dead newborns
- Long term : reduced feed intake, milk fat test, milk yield

Pigs
- Reproductive failure

Poultry
- Reduced hatchability, prolonged hatching time, retardation in absorption of yolk sac, reduced embryo development
- Poor egg production and size
- Growth of abnormally long, lacy feathers
- Male testes remain small and free of spermatozoa, comb reduces, lost plumage, moulting inhibited

Horses
- Reduced libido and poor semen quality
- Goitre in young at birth
- Weakness, persistent hypothermia, respiratory distress, high neonatal mortality
- Increased susceptibility to infectious disease and respiratory infections
- Abnormal oestrus cycles
- Reduced libido and poor semen quality
- Foals born weak and unable to stand and suckle

Toxicity

General
- Increases metabolic rate
- The ability to withstand excess depends on the species
- Recovery often rapid once excess removed
- Most species suffer abortions
- Not usually seen as the animals can tolerate 50 x requirements

Ruminants
- Cows with an intake above 50 mg/day, which is about 5-mg/kg dietary DM
- Young stock more sensitive than lactation cows
- Depressed appetite
- Dull, dry coats and skin
- Increased nasal and ocular discharge
- Increased salivation & excessive tears
- Coughing & difficulty swallowing
- Reduced feed intake and production
- Impaired immune response and reproductive efficiency
- Decreased fertility
- Reduced milk output
- Weak young often with pneumonia

Pigs
- >800ppm depresses growth rate, feed intake, haemoglobin level ad liver Fe concentration

Poultry
- Reduced egg production and size, poor hatchability and embryo survival
- Weak chicks of low viability

Rabbits
- Increased prenatal mortality

Horses
- Mare on excessive I, 3 months prior to foaling results in Hypothyroidism of foals: stillbirths, long hair, weakness, alopecia, contracted tendons, limb abnormalities (long bones),poor muscle development
- Protruding jaws, parrot mouth
- Mares receiving >40mg/d results in foals with goitre

Iodine

11e

| I | Atomic Wt 126.9 | Atomic No 53 | Micro Mineral |

Toxicity

Humans
- Humans are very sensitive to thyrotoxicosis and therefore the diet of cows should not contain more than 0.5 mg/kg or there will be exessive levels in meat and milk. Maximum inclusion limits are enforced.

Maximum Dietary Tolerable Level (ppm DM)

Ruminants *	50	Poultry	300
Sheep	50	Horses	5
Pigs	400		

Notes
* May be an undesirable level in milk

Antagonists

- Goitrogens: interfere with the synthesis/secretion of thyroid hormones and cause hypothyroidism.
  There are 2 basic types:
  Cyanogenic goitrogens
  – reduce iodide uptake.
  Cyanogenic glucosides
  – alter iodide transport across the thyroid follicular cell membrane and therefore reduce the iodide retained
- Cyanogenic glucosides are found in corn/maize, millet, raw soybeans, sugar beet pulp, sweet potato, white clover
- Progoitrins and goitrins are found in various feedstuffs. They interfere with hormone synthesis and a maximum level should be applied
- Found in Cabbage, Kale, Mustard, Rape, Turnips, Swede.
- Onions contain aliphatic disulphides which also inhibit hormone production at higher levels
- Depressants: As, F, Ca, Co, Mn, high K, high dietary N
- Selenium deficiency will influence thyroid hormone metabolism
- High I intake with Se deficiency may initiate thyroid tissue damage from low GSHPx activity
- Both low and high manganese and cobalt may reduce the iodine content of the thyroid gland
- In cows, high calcium levels may reduce the milk iodine content
- Fe deficiency anaemia impairs thyroid metabolism

Main Supplements

Source	Element %	Relative Bio-avail.	Comments
Potassium Iodide (KI)	69-75	High	Water soluble
Sodium Iodide (NaI)	85	High	Store away from other minerals
Calcium Iodate hexahydrate	50	High	
Calcium Iodate anhydrous	65	High	Reduce exposure to light
Pentacalcium orthiperiodate	28	High	Used in feed blocks
Diiodthymol			
Diiodosalicyclic acid		Low	
Iodised salt	0.00756		Water soluble
Cuprous iodide	66.6	High	
Seaweed	0.4-0.6		
Ethyleneediamine dihydroiodide (EDDI)	80	High	# Low water solubility Used in feed blocks

Notes
Supplementation is often given to animals in salt licks, blocks, boluses, drenches or the diet.
Some liberation of free iodine when mixed with Cu, Zn Fe sulphates with heat and moisture. Prolonged storage, heat, moisture, sunlight cause release on free iodine.

Iodine

11f

| I | Atomic Wt 126.9 | Atomic No 53 | Micro Mineral |

Feed Name	mg/kg DM	Feed Name	mg/kg DM
Alfalfa meal	0.15	Molasses -cane	2.10
Bakery waste	-	Oat groats	0.12
Barley grain	0.17	Oat middlings	-
Bean field	-	Oatfeed	-
Blood meal	0.88	Oats grain	0.12
Brewers grains	0.07	Palm kernel exp	1.20
Buckwheat grain	-	Peas	0.16
Buttermilk dehyd.(cattle)	0.21	Potato dried	0.22
Casein dehyd. (cattle)	-	Rape ext (mech)	0.44
Cassava tubers dehy	-	Rice bran	-
Citrus pulp dried	0.10	Rice grain	0.04
Copra meal	1.20	Rye grain	0.11
Cottonseed Whole	-	Safflower ext. solv.	-
Cottonseed meal	0.14	Sesame ext mech	0.16
Distillers grains - wheat	0.11	Silage alfalfa	-
Distillers grains maize	0.24	Silage grass	-
Distillers grains- barley	0.04	Silage maize	0.10
Fishmeal (Sth Am)	-	Silage sorghum	-
Grass bluegrass	-	Silage wholecrop	-
Grass alfalfa	-	Sorghum grain	0.04
Grass bermuda	0.12	Soya ext.solv	0.22
Grass clover	-	Soya flour	-
Grass extensive	0.50	Soya hipro	0.23
Grass kikuyu	0.26	Straw barley	0.07
Grass timothy	-	Straw oat	0.06
Groundnut ext	0.10	Straw wheat	0.06
Hay alfalfa	0.17	Sugar beet pulp (dehyd)	2.00
Hay bluegrass	-	Sugar beet pulp (mol)	0.51
Hay clover	0.25	Sunflower ext	0.68
Hay eragostus	-	Triticale grain	-
Hominy feed	0.10	Wheat (caustic)	0.10
Linseed meal (mech ext)	0.07	Wheat bran	0.09
Maize bran	-	Wheat feed	-
Maize germ ext(sol)	-	Wheat germ ext.	-
Maize gluten 20	0.10	Wheat grain	0.10
Maize gluten 60	0.78	Whey low lactose	10.55
Maize grain	0.02	Whey(cattle dehyd)	-
Malt culms	-	Yeast (brewers dehyd)	0.38
Milk (cattle-dehyd)	0.67	Yeast (torula dehyd)	2.69
Milk skimmed	1.06	....	
Millet grain	-	....	
Molasses -beet	1.40	....	

CONTEXT

www.contextbookshop.com

Iron

12a

| Fe | Atomic Wt 55.85 | Atomic No 26 | Micro Mineral |

Introduction
- A heavy metal, when pure is a dark, silvery-gray
- Very reactive element that oxidises (rusts) very easily
- Naturally magnetic
- Second most abundant metal in earth's crust, about 5%
- The principal ores of iron are Hematite, (Iron Oxide, 70% iron) and Magnetite, (72 % iron). Taconite is a low-grade iron ore, containing up to 30% Magnetite and Hematite
- Iron ore is the raw material used to make pig iron, which is one of the main raw materials to make steel
- Powdered iron is used in:
  -Metallurgy products, magnets, high-frequency cores, auto parts, catalyst, radioactive iron (iron 59),
  -Medicine, tracer element in biochemical and metallurgical research.
- Iron blue is used in:
  -Paints, printing inks, plastics, cosmetics (eye shadow), artist colours, laundry blue, paper dyeing, fertiliser ingredient, baked enamel finishes for autos and appliances, industrial finishes
- Black iron oxide is used in
  -Pigments, polishing compounds, metallurgy, medicine, magnetic inks, in ferrites for electronics industry
- Name 'Iron' comes from an Old English word *'isaern'* which itself can be traced back to a Celtic word, *'isarnon'*
- Iron is essential to animal life and necessary for the health of plants
- Found in the body at only 50-70ppm. Level varies from birth to maturity
- Mainly found as an organic complex with proteins
- About 60% of body Fe is in blood, as an essential part of haemoglobin (gives blood its red colour) and 7% in myoglobin
- Iron deficiency is one of the most common deficiency diseases of indoor reared baby pigs and humans

Key Natural Sources
- Content highly variable in most feed materials
- Plant uptake is determined by species, soil and climatic conditions
- Acid soil conditions encourage availability and plant uptake
- Large concentrations of iron are found in green leaves and in seed husks
- Feed levels will vary with processing procedure
- Soil is a high source and can contaminate feeds
- Most forages contain levels in excess of requirements for herbivores
- Milk is a very poor source
- Legumes and oilseeds are richer than cereals that are low in iron
- Bioavailability from cereal grains and oilseeds reduced by phytate
- Meat meals and fishmeals are rich sources of Fe
- Organic Fe usually less soluble and poorly absorbed compared to inorganic sources
- Drinking water can be an important source of Fe

Function
Involved in various metabolic processes.
- A component of proteins e.g.
  -Myoglobin, hemoglobin, haemerythrin, and haemocyanin involved in oxygen transport
  -Serum protein ferritin, transferrin involved in transfer to parts of the body Lactoferrin, an Fe containing glycoprotein secreted by mammary cells, has an antibiotic agent role in the gland
  -Uteroferrin, a progesterone-inducible protein isolated in sow placenta thought to have a role in transfer of Fe to foetal piglet
- A component of enzymes e.g.
  -Oxidases, catalase, reductases, nitrogenases, hydroxylases, hydrogenases, super dismutase, arginase, ferredoxin, peroxidases, cytochromes, phosphatase
- A component of non proteins e.g.
  -Siderophores for metal storage and transport

CONTEXT www.contextbookshop.com

Iron

12b

| Fe | Atomic Wt 55.85 | Atomic No 26 | Micro Mineral |

Benefits
- Cellular respiration
- Metabolism
- Activation of oxygen and electron transport
- Immune system
- Iron oxide is often used to colour mineral supplements

Absorption
Affected by:
- Age, Fe status, state of health, conditions of the gastrointestinal tract, amount and form of Fe ingested, amount and proportion of other components of diet (organic and inorganic), and genetic control (at least for excess absorption)
- Found in feed in two forms, heme (organic) and nonheme (inorganic), both poorly absorbed, approximately 10%
- Absorption occurs throughout GI tract, mainly in duodenum and jejenum
- Fe absorbed in ferrous state, in ferric form in feed and in combination with organic compounds
- Absorption enhanced by more acid conditions (e.g.. Gastric HCL)
- Inorganic forms, absorption enhanced by the presence of Vitamin C (ascorbic acid), cysteine, lysine, histidine that help convert Ferric Fe^{3+} to ferrous Fe^{2+}
- The organic portion is more freely absorbed and is independent of Vitamin C or iron chelating agents.
- Iron absorption controlled by the Fe concentration in the mucosal epithelial cells of the duodenum

Metabolism
- Iron, taken into mucosal cells is converted to ferritin
- Mucosal ferritin supplies ferrous iron (Fe^{2+}) to blood plasma
- Here it is oxidised to the ferric (Fe^{3+}) state and immediately combines with a protein (transferrin) and transported throughout the body
- Transferrin accepts Fe from intestinal tract and released from storage sites and from haemoglobin destruction
- Transferrin transports Fe (70% of plasma Fe) to the bone marrow for haemoglobin synthesis, to the placenta for foetal needs and cells for Fe-containing enzymes.
- Storage forms, ferritin and hemosiderin, reflect animals Fe status, and mainly found in the liver, reticuloendothelial cells and bone marrow
- Hemosiderin is predominant form at high tissue levels and ferritin predominates at lower levels
- Synthesis of heme is dependant on sufficient Fe and Cu. (Cu dependant enzyme, cytochrome oxidase reduces Fe^{3+} to Fe^{2+})
- Cellular Fe homeostasis is through controlled synthesis of several proteins (iron regulatory proteins, IRPs) involved in movement, storage and utilisation of Fe
- Storage Fe can be reduced from ferritin to ferrous form mainly by riboflavin, vitamin C, glutathione or cysteine. Mobilisation of Fe from stores also requires a Cu-containing enzyme
- Young animals are born with a store of iron in their livers, influenced by maternal diet, particularly in late gestation
- Milk is low in iron and so a long period on milk diets lowers levels

Excretion
- Not readily lost from body except through haemorrhage
- The body conserves absorbed iron by reusing it
- Fe released from haemoglobin during red blood cell breakdown is carried to liver and secreted as bile
- Most bile Fe is reabsorbed and used again to form haemoglobin
- Most is lost in the faeces and the urine, with some loss through sweat, hair and nails by desquamation (shedding of cells)
- Most faecal Fe is unabsorbed iron

CONTEXT

www.contextbookshop.com

Iron

Fe | Atomic Wt 55.85 | Atomic No 26 | **Micro Mineral**

Requirement /Allowances (mg/kg DM)

Rums	NRC	Pigs	NRC	Poultry	NRC	Others	NRC
Calf	(a)	Creep	100	Chick	80	Dog	(b)
Dairy	12-18	Weaner	100	Broiler	80	Cat	80
Beef	50	Grower	80	Breeder	45	Horse	40-50
Heifer	31-43	Finisher	40-60	Layer	45	Fish	30-150
Sheep	30-50	Sow/Boar	80	Turkey	60-80		

Rums	Typical	Pigs	Typical	Poultry	Typical	Others	Typical
Calf	30-100	Creep	60-200	Chick	10-80	Dog	80
Dairy	40-100	Weaner	150-250	Broiler	20-40	Cat	80
Beef	50	Grower	100-200	Breeder	20-40	Horse	50-80
Heifer	75	Finisher	100-170	Layer	10-80	Fish	50
Sheep	40-100	Sow/Boar	100-200	Turkey	20-80	Rabbits	40

Notes
(a) Milk replacer 30-50 mg/kg DM
(b) 0.65-1.74 mg/kg BW

- EU regulation: maximum inclusion in feedstuffs for producing animals: Poultry- 500mg/kg, Pig– 750mg/kg, Beef, Dairy -750mg/kg;
- Veal diets at 25-30mg/kg DM, iron must be in readily available form. Level provides sufficient haemoglobin for normal appetite, growth and oxygen transport. Higher level required by some breeds to allow maximum production of haemoglobin and myoglobin in rapidly growing calves
- The iron requirement increases during pregnancy and as the foetus increases in size
- Laying hens have large metabolic demand for iron, as the egg contains high levels. (1.5mg Fe)
- Bleeding increases the requirement as does heavy infestation with intestinal parasites

Adequate Status

Species	Liver	Serum Fe	Serum Hb	Serum Ferritin	Kidney
Cattle	45-300	130-250	9-14	30-50	30-150
Sheep	30-300	166-222	8-16		30-200
Pigs	100-200	80-180	10-16	55-65	
Poultry	60-300		10-15		45-100
Horses	100-300	75-140	11-19		35-150
Dogs	100-300	94-122	12-18		75-260
Units	ppm WW	μg/dl DM	g/dl DM	ng/ml DM	ppm WW

Note
For cattle the level in milk is typically 0.2-6.3 mg/l DM

- Haemoglobin and hemocrit values, serum Fe, total Fe-binding capacity, % transferrin saturation can be used to assess status.
- Haemoglobin and hemocrit levels are not sensitive to early Fe deficiency because only occur when storage Fe is severely depleted.
- Serum Fe and % transferrin saturation are early indicators of Fe deficiency
- Serum ferritin relates to amount of Fe stores.
- Infectious diseases which cause fever, reduce serum Fe.

Iron

12d

| Fe | Atomic Wt 55.85 | Atomic No 26 | Micro Mineral |

Deficiency

General
Primary: Nutritional anaemia (low red blood cell count) paleness of mucous membrane
- Changes to skin, hair condition
- Diarrhoea/scours
- Fatigue
- Retarded growth in foetus and youngstock
- Poor growth
- Reduced appetite (pica)
- Increased susceptibility to disease
- Deficiency primarily during rapid growth and suckling phases, also in pregnancy and following accidental haemorrhage

Pigs
- Common in housed piglets as sow's milk is low in iron, low stores and very rapid early growth rate
- Hypochromic-microcytic anaemia. Blood haemoglobin fall from 10g/dl to as low as 4g/dl within 2-4 weeks of birth
- Chronic: rough hair/coat, wrinkled skin, pale mucous membranes, listless, drooping head and upper eyelids, ears and tail hang limp
- Acute: laboured breathing following exercise, low resistance to disease, respiratory problems and enteritis
- Older pigs, on high Cu diets to promote growth, develop anaemia if Fe not over -supplemented

Ruminants
- Most likely in young animals as milk is low in Fe. See symptoms above
- Rarely seen in grazing animals as forages/ingested soil provide adequate levels

Poultry
- Rarely seen on practical diets
- Lower growth rates and reduced haemoglobin
- Poor feathering and impaired plumage pigmentation
- Breeding hens - embryonic mortality during 9-15 days incubation

Veal
- Show a light coloured muscle due to low myoglobin as a result of reduced dietary iron without other foodstuffs

Horses
- Microcytic and hypochromic anaemia
- More susceptible to stress factors and diseases, reduced growth, scours and pneumonia
- Heavy parasitism may exasperate deficiency

Toxicity

General
- Free iron is very toxic so it is normally transported in combination with protein. Under normal circumstances the amount of iron in the plasma is sufficient to bind only 1/3 of the transferrin and the remaining 2/3 remains as unbound reserve.
- When the level of iron exceeds the binding capacity of the transferrin, iron toxaemia occurs
  Most animals have a high tolerance
- Chronic signs: Reduced feed intake, weight gain and feed efficiency
- Acute signs: anorexia, diarrhoea, hypothermia, metabolic acidosis, death
- Ruminants reduced DM digestibility
- Reduced milk output

Pigs
- Injectible Fe at 100mg is toxic to piglets from vitamin E-Se deficient sows
- Shivering, incoordination, titanic convulsions

Maximum Dietary Tolerable Level (ppm DM)

Cattle	1000	Poultry	1000
Sheep	500	Horses	500
Pigs	3000		

CONTEXT   www.contextbookshop.com

Iron

12e

| Fe | Atomic Wt 55.85 | Atomic No 26 | Micro Mineral |

Antagonists
- Absorption is impaired by presence of: organic acids, feeds high in inorganic iron, some phosphates, phytates, polyphenols, oxalates, gossypol (toxic component of cottonseed meal), tannic acid, aflatoxin
- High dietary levels of Cu, Mn, Pb and Cd decrease absorption by competing for absorption sites in intestinal mucosa
- Excess Fe affects availability of P, Vitamin A and Cu in the diet
- Excess Fe increases need for Na and K
- Excess Fe can induce Co, Cu, Mn, Se, and Zn deficiency
- High tissue Fe inhibits use of tissue Se for glutathione peroxidase and reduces deposition rate of Vitamin A in the livers of young animals
- Dietary imbalances of zinc can affect absorption and use of iron and copper
- High dietary Mn (1000ppm DM cattle) reduces liver and pancreas Fe, if dietary Fe is normal

Synergy
- Cu, Co, Mn and Vitamin C are required for iron to be used by the animal. Absorption of iron is increased by ascorbic acid (Vit C), tocopherol (Vit E), fructose and sulphur containing amino acids.
- Fe deficient diet increases Cd, Co, Mn, Pb, Zn absorption

Main Supplements

Source	Element %	Relative Bio-avail.	Comments
Ferrous ammonium citrate			
Ferric chloride			
Ferric ortho phosphate			
Ferric pyrophosphate			
Ferrous chloride			
Ferrous fumarate			
Ferrous gluconate			
Ferrous succinate			
Ferrous sulphate	20-30	high	High water solubility High hygroscopicity
Iron carbonate	36-42	low	Not avail to chicks
Iron oxide	46-60	unavail	Used as a colourant
Iron methionine	15	High	
Iron dextran injection			
Iron peptonate			

CONTEXT www.contextbookshop.com

Iron

| Fe | Atomic Wt 55.85 | Atomic No 26 | Micro Mineral |

Feed Name	mg/kg DM	Feed Name	mg/kg DM
Alfalfa meal	255.0	Molasses - cane	265.0
Bakery waste	31.0	Oat groats	54.4
Barley grain	85.0	Oat middlings	420.0
Bean field	85.0	Oatfeed	44.4
Blood meal	4064.0	Oats grain	75.0
Brewers grains	266.0	Palm kernel exp	483.5
Buckwheat grain	50.0	Peas	62.0
Buttermilk dehyd.(cattle)	9.0	Potato dried	74.0
Casein dehyd. (cattle)	15.0	Rape ext (mech)	244.4
Cassava tubers dehy	10.0	Rice bran	210.0
Citrus pulp dried	378.0	Rice grain	57.0
Copra meal	472.0	Rye grain	69.0
Cottonseed whole	151.6	Safflower ext. solv.	537.0
Cottonseed meal	197.0	Sesame ext mech	100.0
Distillers grains - wheat	351.0	Silage alfalfa	305.0
Distillers grains maize	241.1	Silage grass	175.0
Distillers grains- barley	220.0	Silage maize	165.0
Fishmeal (Sth Am)	392.3	Silage sorghum	285.0
Grass bluegrass	255.0	Silage wholecrop	86.0
Grass alfalfa	286.0	Sorghum grain	51.0
Grass bermuda	290.0	Soya ext.solv	185.0
Grass clover	307.0	Soya flour	-
Grass extensive	122.0	Soya hipro	170.5
Grass kikuyu	173.0	Straw barley	200.0
Grass timothy	184.0	Straw oat	185.0
Groundnut ext		Straw wheat	160.0
Hay alfalfa	200.0	Sugar beet pulp (dehyd)	329.0
Hay bluegrass	293.0	Sugar beet pulp (mol)	168.5
Hay Clover	184.0	Sunflower ext	339.0
Hay eragostus	-	Triticale grain	27.5
Hominy feed	75.0	Wheat (caustic)	58.1
Linseed meal (mech ext)	212.0	Wheat bran	208.0
Maize bran	287.0	Wheat feed	84.3
Maize germ ext(sol)	354.0	Wheat germ ext.	58.0
Maize gluten 20	28.0	Wheat grain	58.1
Maize gluten 60	333.3	Whey low lactose	262.0
Maize grain	216.0	Whey (cattle dehyd)	181.0
Malt culms	144.0	Yeast (brewers dehyd)	117.0
Milk (cattle-dehyd)	10.0	Yeast (torula dehyd)	126.0
Milk skimmed	8.7		
Millet grain	68.5		
Molasses -beet	140.0		

Bioplex® Iron from Alltech®

| Fe | Atomic Wt 55.85 | Atomic No 26 | Micro Mineral |

Introduction
- Bioplex® Iron is an organic trace mineral proteinate for use in livestock feeds.

Concentrations Available

Product	Guaranteed Analysis	Ingredients
Bioplex® Fe 10%*	Minimum 10% Iron	Iron proteinate
Bioplex® Fe 15%*	Minimum 15% Iron	Iron proteinate

Not all mineral concentrations are available in every country. Contact your local Alltech representative for details.

Physical Characteristics

Appearance
Bioplex® Iron is a dark brown coloured, free flowing powder with no discernible odour.

Storage
Bioplex® Iron should be stored in a closed container in a cool, dark area. Shelf life under these conditions is 36 months.

Packaging
Bioplex® Iron is available in 25 kg bags.

Inclusion Rates of Bioplex® Minerals
- Every species requires a different inclusion rate. Also inclusion depends on the motivation for organic supplementation. There are two issues to be examined.
- Organic supplementation can be for performance reasons, here research shows that supplementation with Bioplex® minerals at rates between 20-50% of inorganic minerals provides superior performance effects.
- Supplementation may be driven by the requirement to meet environmental regulations. Bioplex® Iron can be used as a complete replacement of inorganic salts at a significantly lower inclusion level.

For inclusion recommendations for your region and other details, contact your local Alltech representative.

Benefits (Conditions responsive to improved Iron status)
- Iron is key to the transport of oxygen in the blood
- An essential part of many enzymes including catalases, cytochromes and peroxidases
- Promotes immune function
- Involved in cellular respiration and metabolism

Some Symptoms of Iron Deficiency
- The main iron deficiency is anaemia, which leads to fatigue and failure to thrive
- Iron deficiency can also cause abnormalities in epithelial tissues
- Increased susceptibility to disease

www.alltech.com

Lead

13a

Pb | Atomic Wt 207.19 | Atomic No 82 | **Highly Toxic Mineral**

Introduction
- A soft, blue-grey, metallic element
- A heavy metal, easily melted and which readily forms alloys with many other metals
- Name 'Lead' comes from Latin word *'Plumbum'*, and from its use in making water pipes, 'plumber and 'plumbing' were derived
- Historically it was a component of: lead batteries, paint ('lead free paint' may contain up to 1% Pb) petrol/fuel and passed into the environment through emissions, roofing asphalt, window putty, waste engine oil, linoleum, solder, golf balls, lead shot, weights
- Pb absorbs radiation from radioactive isotopes
- Prevalent in ambient air, dust, industrial gasoline, inks and paints
- The uptake by plants is limited and affected by soil pH but deposition on leaves is common from vehicle exhaust fumes
- Levels increase in forage during winter months
- It is also found in rainwater in areas of coal-fired power stations and in other watercourses that come into contact with lead pipes
- Acts as a cumulative poison to livestock but new research with rats, on very pure sources, suggest it as an essential trace mineral involved with blood and growth functions – no such link has been found in livestock, yet

Key Natural Sources
Typical levels:

Grass	4.3 mg/kg
Grass Silage	4.2 mg/kg
Wheat	1.3 mg/kg
Barley	1.0 mg/kg
White fish meal	3.7 mg/kg
Potatoes	3.2 mg/kg

Function
- Inhibits enzymes dependent upon the presence of free sulfhydryl groups for activity

Absorption
- Absorbed from food into blood and carried to bones, kidney, liver and hair.
  Young 15- 50%
  Adults 5-10%
- More lead is absorbed from inorganic salts than from organic
- Also absorbed through the respiratory tract and the skin
- Absorption increased in fasted rather than fed subjects

Metabolism
- Approx. 90% retained largely sequestered in the skeleton and relatively immobile
- Excess accumulates in kidneys and liver (normal kidney level ruminants 20 mg/kg fresh, versus 50 mg/kg in animals which have died from lead poisoning)
- Limited levels in milk and muscle
- Can be transferred readily across placenta in humans, rats and goats

Excretion
- Via bile and urine and is very slow

Requirement /Allowances (mg/kg DM)
- Inclusion levels should be as low as possible (<5mg/kg DM)

CONTEXT www.contextbookshop.com

Lead

13b

| Pb | Atomic Wt 207.19 | Atomic No 82 | Highly Toxic Mineral |

Deficiency

General
- Growth reduction and anaemia (deficiency is unlikely to occur)

Toxicity

General
- Affects heme synthesis and protein synthesis
- Anaemia
- Colic, pain in intestines
- Impaired neurologic functions leading to irritability and blindness
- Interference with metal-dependent enzymes functions
- Abortions (related to Ca, lead ions replace Ca in cells)
- Reduced antibody synthesis, increases susceptibility to disease

Ruminants
- More susceptible. Symptoms include anorexia, fatigue, depression, hyperexcitability, diarrhoea or constipation (dark grey faeces), salivation, loss of weight and ataxia
- Reduces rate of fermentation by ruminal microorganisms
- Acute cases = death in 2-3 days
- Calves- difficulty swallowing or suckling

Horses
- Paryngeal and laryngeal paralysis (roaring), colic, diarrhoea. Reduced immune response (probably induced) Joint stiffness

Pigs
- Extremely resistant to Pb poisoning and toxicosis rare
- Inappetance, listlessness, muscle tremors, incoordination, increased respiratory rate, enlarged carpal joints, mild clonic seizures

Poultry
- Drowsiness, thirst, weakness, anorexia, diarrhoea, anaemia, crop-stasis. Reduced immune response, growth rate and egg production
- Soft-shelled eggs

Dogs
- Can be confused with canine distemper. Affects gastrointestinal and nervous system. Blue lead-line between gum and tooth has been seen in dogs and possibly osteoporosis symptoms due to weakening effect of lead in the bones

Maximum Dietary Tolerable Level (ppm DM)

Ruminants	30	Rabbits	30
Pigs	30	Horses	30
Poultry	30		

Notes
Based on human food residue considerations

Interrelationships
- Affects Cu, Fe, Se metabolism

Antagonists
- Absorption is affected by calcium, phosphorus, iron, copper, zinc, manganese and ascorbic acid, fat, protein levels in the feed

Synergy
- Fe, S, Zn, Vitamin C and E help protect against Pb toxicity

Main sources
- Legal maximum in feedstuffs 5 mg/kg
- Mn Oxide for animal feeds must have no more than max.100ppm Pb

CONTEXT   www.contextbookshop.com

Lead

13c

| Pb | Atomic Wt 207.19 | Atomic No 82 | Highly Toxic Mineral |

Feed Name	mg/kg DM	Feed Name	mg/kg DM
Alfalfa meal	-	Molasses - cane	-
Bakery waste	-	Oat groats	-
Barley grain	1.0	Oat middlings	-
Bean field	-	Oatfeed	-
Blood meal	-	Oats grain	-
Brewers grains	-	Palm kernel exp	-
Buckwheat grain	-	Peas	-
Buttermilk dehyd. (cattle)	-	Potato dried	32.0
Casein dehyd. (cattle)	-	Rape ext (mech)	-
Cassava tubers dehy	-	Rice bran	-
Citrus pulp dried	-	Rice grain	-
Copra meal	-	Rye grain	-
Cottonseed whole	-	Safflower ext. solv.	-
Cottonseed meal	-	Sesame ext mech	-
Distillers grains - wheat	-	Silage alfalfa	-
Distillers grains maize	-	Silage grass	4.2
Distillers grains - barley	-	Silage maize	-
Fishmeal (Sth Am)	3.7	Silage sorghum	-
Grass bluegrass	-	Silage wholecrop	-
Grass alfalfa	-	Sorghum grain	-
Grass bermuda	-	Soya ext.solv	-
Grass clover	-	Soya flour	-
Grass extensive	4.3	Soya hipro	-
Grass kikuyu	-	Straw barley	-
Grass timothy	-	Straw oat	-
Groundnut ext	-	Straw wheat	-
Hay alfalfa	-	Sugar beet pulp (dehyd)	-
Hay bluegrass	-	Sugar beet pulp (mol)	-
Hay clover	-	Sunflower Ext	-
Hay eragostus	-	Triticale grain	-
Hominy feed	-	Wheat (caustic)	1.3
Linseed meal (mech ext)	-	Wheat bran	-
Maize bran	-	Wheat feed	-
Maize germ ext(sol)	-	Wheat germ ext.	-
Maize gluten 20	-	Wheat grain	1.3
Maize gluten 60	-	Whey low lactose	-
Maize grain	-	Whey (cattle dehyd)	-
Malt culms	-	Yeast (brewers dehyd)	-
Milk (cattle-dehyd)	-	Yeast (torula dehyd)	-
Milk skimmed	-	....	-
Millet grain	-	....	-
Molasses -beet	-	....	-

CONTEXT

www.contextbookshop.com

Magnesium

14a

| Mg | Atomic Wt 24.31 | Atomic No 12 | Macro Mineral |

Introduction
- Discovered by Sir Humphrey Davy in 1808
- Silvery white metallic element and very light, has countless applications where weight reducing is important
- Very abundant in nature, and is found in important quantities in many rocky minerals, for e.g. dolomite, magnesite, olivine and serpentine
- Found in seawater, underground brines and salty layers
- The third most abundant structural metal in the earth's crust
- Magnesium components are widely used in industry and agriculture
- A critical mineral involved in most biochemical processes in the body 60-70% is found in the skeleton, 30% in soft tissues, body fluids
- Found in all cells and its concentration is higher than any other mineral except K
- A structural component of the skeleton and its functions are closely associated with calcium and phosphorus
- Body contains approximately 0.05%

Key Natural Sources
- Abundant in most common feed stuffs
- Rich sources are wheat bran, wheat meal and rice bran
- Plant proteins are good sources
- Forage levels vary with species, soil type and climate. Legumes tend to be higher than grasses
- Temperature and light may affect forage Mg. Low daily radiation reduces levels
- High fertilisation with K and/or N reduced Mg content of grasses
- Mg concentration greater in stems than leaves
- The availability of forage Mg for ruminants varies from 10-25% and 30-40% in grains and concentrates
- Forage Mg availability appears to increase with maturity & preservation

Function
- Bone and teeth integrity
- Neuromuscular activity
- Involved in the bodies hormonal activity
- Essential for cell respiration
- Enzyme activation
- Carbohydrate, protein, fat metabolism
- Needed for normal insulin sensitivity

Benefits
- Required for growth, repair of body tissues and bone development
- Increases rumen pH and can help increase milk yield and butterfats
- Involved in peptidases in protein digestion
- Appears to have a role in activation of Vitamin D
- Relaxes nerve impulses
- Pigs fed supplement pre-slaughter may be beneficial to meat quality
- Elevated levels improve marbling score in cattle fed fat supplemented diet

Absorption
- Absorption occurs through the digestive tract and endogenous secretions that can be resorbed
- Monogastrics: main site of absorption is the small intestine
- Absorption is by two processes, a carrier mediated system at low concentrations and by simple diffusion at higher concentrations
- Ruminant: main site of absorption is reticulorumen
- Absorption declines with increasing dietary levels
- Mg status of animal alters Mg absorption
- Absorption reduced by high levels of Ca, P, K, Fe, fats, oxalic acid, phytate. Ca and Mg may compete for same absorption sites
- Absorption increased by antibiotics, growth hormone, lactose, protein, vitamin D, ionophores
- Ruminants: absorption improved with more readily degradable carbohydrates
- Rumen pH is important in its influence on solubility. High pH reduces Mg solubility

CONTEXT   www.contextbookshop.com

Magnesium

14b

| Mg | Atomic Wt 24.31 | Atomic No 12 | Macro Mineral |

Metabolism
- The kidneys act to reabsorb wasted magnesium
- Ability for bone mobilisation decreases with age

Excretion
- Major pathway via urine
- Endogenous Mg via faeces from bile, saliva, gastric juices, pancreatic juices, intestinal secretion and intestinal defoliation
- Milking animals' have large loss via milk, particularly early lactation
- Eggshell contains major portion of Mg in eggs

Requirement /Allowances (g/kg DM)

Rums	NRC	Pigs	NRC	Poultry	NRC	Others	NRC
Calf (a)	1.0	Creep	0.4	Chick	0.6	Dog	(b)
Dairy	1.1-2.9	Weaner	0.4	Broiler	0.6	Cat	0.4
Beef	1-2	Grower	0.4	Breeder	0.5	Horse	0.9-1.2
Heifer	1.1	Finisher	0.4	Layer	0.5	Fish	0.4-0.6
Sheep	1.2-1.8	Sow/Boar	0.4	Turkey	0.5	Rabbits	0.3-0.4

Rums	Typical	Pigs	Typical	Poultry	Typical	Others	Typical
Calf	1.0	Creep	0.4	Chick	0.6	Dog	0.4
Dairy	2.2-3.5	Weaner	0.4	Broiler	0.5	Cat	0.4-0.8
Beef	1.5	Grower	0.4	Breeder	0.5	Horse	
Heifer	1.6	Finisher	0.4	Layer	0.45	Fish	
Sheep	2-3	Sow/Boar	0.4	Turkey	0.5	Rabbits	

Notes
(a) 0.7g/kg - milk replacer
(b) 8.2-22 mg/kg bw

Adequate Status

Species	Liver	Kidney	Serum Tl	Milk	Eggshell
Cattle	100-250	50-200	1.8-3.5(a)	10-14	
Sheep	118-200	110-160	2-3.5		
Pigs	150-180	140-180	1.8-3.9		
Poultry			1.3-2.41		0.53-0.64
Horses	130-200	90-200	1.8-3.5		
Dogs	90-200	100-240	1.8-2.7		
Units	ppm WW	ppm WW	mg/100ml DM	mg/100ml DM	%DM

Notes
(a) For calves 2.5

- Serum Mg is good indicator of status in severe deficiency but urine Mg and erythrocytes are better indicators
- Adequate to liberal urine Mg levels: >10mg/dl
- Cattle: Many diseases that result in cell destruction produce transient elevated serum Mg levels

Magnesium

| Mg | Atomic Wt 24.31 | Atomic No 12 | Macro Mineral |

Deficiency

General
- Retarded growth
- Hyper-irritability and tetany
- Anorexia
- Muscular incoordination
- Convulsions
- Excessive calcification of bone and fat tissue

Ruminants
- Hypomagnesaemia, (grass staggers, grass tetany, wheat pasture poisoning) from lowered Mg in the blood (<1.1mg/100ml)
- Affected by forage mineral levels, soil properties, fertlization, season, temperature, animal species, breed, age, stress
- Mature lactating cattle most susceptible
- Tetany in cattle shows signs of depressed appetite, dull, lethargic, stiff movements, staggers, nervousness, muscle twitching, staring eyes, teeth grinding, excess salivation, extreme excitement and convulsions, collapse, thrashing legs
- Rapid treatment is essential or the animal passes into a coma and dies
- Treatment is usually with magnesium sulphate or magnesium lactate
- Tetany in calves is seen when they are fed for a long time on natural milk
- Marginal deficiency: reduced feed intake and performance, possibly reduced cellulose digestion

Pigs
- Low requirements and efficient recycling make deficiency rare
- Young suckling piglets can suffer, as the milk does not supply all the needs – stepping syndrome

Poultry
- Incidence rare
- Reduced growth rate or egg production, neuromuscular hyper-irritability, gasping, convulsions, anaemia, death. Increased incidence of tibial dyschrondroplasia in broiler chick
- Low egg size, weight of shell and Mg content of yolk and shell
- Low shell Mg reduces hatchability

Dogs/Cats
- Anorexia, vomiting, decreased weight gain, hypertension Hyper-irritability, convulsions, soft tissue calcification and enlargement of long bones (dogs). Calcification of the aorta

Horses
- Horses not as susceptible as cattle
- Foals – degeneration in lung, spleen, skeletal muscle and heart

Sheep
- Ewes breathe rapidly, facial muscle tremble, the walk is stiff and awkward, spasm occurs and leg become rigidly extended
- Most common within 4-6 weeks of lambing and can be precipitated by bad weather or stress

Toxicity

General
- Not seen with normal feeds but can be seen when supplements given
- Sustained heart contraction leading to death
- Increase need for phosphorus
- May affect Ca, P and Vit D requirement
- Lethargy, disturbance in locomotion, scours/diarrhoea, reduced feed intake, reduced performance, drowsiness, death
- Excess dietary Mg can lead to Ca urolith-formation, particularly if Ca is low
- Bone Zn declines whilst Mg and P increase

Ruminants
- Diarrhoea and emaciation

Pigs
- Slightly reduced weight gains and feed conversion rate. Elevated serum alkaline phosphatase

Poultry
- Increased mortality and bone deformities, depressed growth rate and bone calcification. Watery faeces

Cats
- Urine pH>6.4, risk of struvite urolithiasis increase as dietary Mg increases

Magnesium

| Mg | Atomic Wt 24.31 | Atomic No 12 | Macro Mineral |

Maximum Dietary Tolerable Level (g/kg DM)

Ruminants	5	Rabbits	3
Pigs	3	Dogs	3
Poultry	3	Horses	3

Antagonists
- Excess Mg affects Ca and P metabolism
- Elevated dietary Ca or P reduce Mg availability

Ruminants:
- High K, Al and N; low Na and fibre; stress, age, high protein, NPN diets and reduced fermentable energy diets

Poultry:
- Increased dietary P or Cl reduce effects of excess Mg
- Salt (NaCl) deficiency can lead to elevated serum Mg
- Excessive dietary fat increases Mg requirement

Pigs:
- High dietary Mn depresses heart concentrations of Mg leads to convulsive seizures and death

Synergy
- Lactose enhances Mg absorption
- High dietary F and Mg are synergistic in reducing growth rate and bone ash
- Sheep: monensin enhances Mg absorption and retention
- Mg can help activate enzymes but is also capable of inactivating certain enzymes

Main Supplements
- Most important for ruminants for prevention and cure of hypomagnesemia
- Most practical poultry and swine diets provide sufficient Mg to meet requirements
- Methods of providing magnesium include:
  - Mg fertilisation of pastures and Hi-Mg forage varieties
  - Foliar Mg application, (fine calcined magnesite)
  - Oral Mg supplementation
  - Mg supplemented water
  (Important to assess advantages and disadvantages for production facility)

Main Supplements

Source	Element %	Relative Bio-avail.	Comments
Dolomitic/Mg limestone	10-13	low	
Mg limestone	3-9		
Magnesium carbonate	21-28	high	
Magnesium hydroxide	30-40	med	Water soluble Has laxative Effect
Magnesium phosphate	24		
Mg chloride	12	high	Water soluble
Magnesium oxide *	54-60	med	
Magnesium sulphate (Epsom salts)	9.6-17	high	Water soluble Has laxative effect
Potassium/Mg sulphate	11	high	

Notes
* most commonly used considering cost per unit biological availability. Particle size, processing temperature and physiological state of animal influence availability

Magnesium

14e

| Mg | Atomic Wt 24.31 | Atomic No 12 | Macro Mineral |

Feed Name	g/kg DM	Feed Name	g/kg DM
Alfalfa meal	2.4	Molasses - cane	4.6
Bakery waste	2.6	Oat groats	1.3
Barley grain	1.3	Oat middlings	1.5
Bean field	1.6	Oatfeed	0.7
Blood meal	2.4	Oats grain	2.0
Brewers grains	1.7	Palm kernel exp	2.8
Buckwheat grain	1.2	Peas	1.3
Buttermilk dehyd.(cattle)	5.2	Potato dried	1.4
Casein dehyd. (cattle)	0.1	Rape ext (mech)	4.4
Cassava tubers dehy	1.1	Rice bran	10.4
Citrus pulp dried	1.5	Rice grain	1.5
Copra meal	3.3	Rye grain	1.4
Cottonseed whole	3.5	Safflower ext. solv.	3.7
Cottonseed meal	5.8	Sesame ext mech	5.0
Distillers grains - wheat	2.8	Silage alfalfa	3.8
Distillers grains maize	0.8	Silage grass	1.3
Distillers grains- barley	3.3	Silage maize	1.0
Fishmeal (Sth Am)	2.2	Silage sorghum	2.9
Grass bluegrass	1.7	Silage wholecrop	0.8
Grass alfalfa	2.7	Sorghum grain	1.8
Grass bermuda	1.7	Soya ext.solv	2.9
Grass clover	4.8	Soya flour	3.6
Grass extensive	1.6	Soya hipro	3.2
Grass kikuyu	3.0	Straw barley	1.1
Grass timothy	1.4	Straw oat	1.2
Groundnut ext	3.5	Straw wheat	1.0
Hay alfalfa	3.3	Sugar beet pulp (dehyd)	2.7
Hay bluegrass	1.6	Sugar beet pulp (mol)	1.7
Hay clover	4.3	Sunflower ext	6.2
Hay eragostus	1.9	Triticale grain	1.1
Hominy feed	2.2	Wheat (caustic)	1.2
Linseed meal (mech ext)	5.8	Wheat bran	5.7
Maize bran	2.1	Wheat feed	2.3
Maize germ ext(sol)	2.2	Wheat germ ext.	2.8
Maize gluten 20	4.0	Wheat grain	1.2
Maize gluten 60	0.7	Whey low lactose	2.3
Maize grain	1.3	Whey (cattle dehyd)	1.4
Malt culms	1.6	Yeast (brewers dehyd)	2.7
Milk (cattle-dehyd)	1.0	Yeast (torula dehyd)	1.8
Milk skimmed	1.3		
Millet grain	1.8		
Molasses -beet	3.0		

CONTEXT www.contextbookshop.com

Bioplex® Magnesium from Alltech

| Mg | Atomic Wt 24.31 | Atomic No 12 | Macro Mineral |

Introduction
- Bioplex® Magnesium is an organic trace mineral proteinate for use in livestock feeds.

Concentrations Available

Product	Guaranteed Analysis	Ingredients
Bioplex® Mg 10%*	Min. 10% Magnesium	Magnesium proteinate

*Not all mineral concentrations are available in every country. Contact your local Alltech representative for details.

Physical Characteristics

Appearance
Bioplex® Magnesium is a cream coloured, free flowing powder with no discernible odour.

Storage
Bioplex® Magnesium should be stored in a closed container in a cool, dry and dark area. Open bags should be resealed. Shelf life under these conditions is 36 months.

Packaging
Bioplex® Magnesium is available in 25 kg bags.

Inclusion Rates of Bioplex® Minerals
- Bioplex® Magnesium is used for specific issues. Consequently, use rates vary on species and specific requirements.

For inclusion recommendations for your region and other details, contact your local Alltech representative.

Benefits (Conditions responsive to improved Magnesium status)
- Magnesium is required for bone and teeth formation
- Maintains normal muscle and nerve function
- Involved in energy metabolism

Some Symptoms of Magnesium Deficiency
- Nervousness
- Muscle spasms
- Reduced appetite and performance

www.alltech.com

Manganese

15a

| Mn | Atomic Wt 54.94 | Atomic No 25 | Micro Mineral |

Introduction
- A gray-white heavy metal with a pinkish tinge, and a very brittle but hard metallic element
- A reactive element that easily combines with ions in water and air; it is not found naturally in its elemental state
- Can exist in 11 oxidation states e.g. oxides, hydroxides, carbonates and silicates
- In the Earth, manganese is found in a number of minerals of different chemical and physical properties, but is never found as a free metal in nature. The most important mineral is pyrolusite, because it is the main ore mineral for manganese
- Name 'Manganese' comes from the Latin word *magnes* which means magnet. When manganese is alloyed with other metals like aluminum, copper and antimony, the end product is magnetic
- Over 80% of the known world manganese resources are found in South Africa and Ukraine
- Important in steel industry as steel becomes harder when it is alloyed with manganese
- Manganese dioxide is used to: manufacture ferroalloys; manufacture dry cell batteries (it's a depolariser); to "decolorise" glass; to prepare some chemicals, like oxygen and chlorine; and to dry black paints
- An essential element for plants/animals
- Present in the body at approx 0.25mg/kg of body weight
- 25% of total body Mn is found in the skeleton
- Bones, liver, pancreas, and kidney have relatively high levels
- Especially concentrated in the reproductive organs
- Has similarity to Iron (Fe) in its chemical properties

Key Natural Sources
Plants variable and dependent on:
- Fertiliser application
- Plant species. (cereal and legume crops-lower DM levels than forages)
- Soil pH (more alkaline soil, the lower the manganese level)
- Soil type (sandstone soils have the lowest Mn content)

Function
- Essential for the functioning of the brain and nervous systems
- An enzyme activator in the metabolism of carbohydrates, proteins, fats and nucleic acids
- A constituent of metalloenzymes.
- Essential for development of organic matrix of bone (mucopolysaccharides)
- Maintains bone mineralisation
- Involved in redox processes
- Co factor of glycoproteins (ie) Blood clotting
- Involved in:
  - the biosynthesis of choline
  - cholesterol synthesis
  - insulin activity
  - production of thyroxine
- Maintains normal central nervous system function
- Enables body to use thiamin (Vitamin B1) and Vitamin E

Benefits
Has a direct effect on:
- Connective tissue formation and growth
- Formation of cartilage
- Mucous production
- Bone development
- Egg production, shell quality, hatchability
- Tissue respiration, (Important to the Krebs cycle)
- Reproduction (corpus luteum functioning)
- Blood formation
- Production of hormones
- Immune function
- Brain function
- Aids use of fats and can help prevent fatty degeneration of the liver

Absorption
Absorbed along in the small intestine
- Poorly absorbed versus other minerals
- Approximately 0.1-0.5% of ingested is absorbed (dependent on species, age etc.
- P, Ca, Mg, Fe can influence absoprtion
- Phytate levels and fibre can reduce absorption by formation of complexes

CONTEXT www.contextbookshop.com

Manganese

15b

| Mn | Atomic Wt 54.94 | Atomic No 25 | Micro Mineral |

Metabolism
- No appreciable stores in the body
- Widley distributed throughout the body but in low concentrations, higher levels in bone, liver, kidney and pancreas
- Tissue levels remain relatively constant over a wide range of dietary intake
- Mn is bound to transferrin in the liver and released into the circulation for transport to the tissues
- Manganese accumulates in the liver (up to a certain level) at a rate proportional to dietary supply
- Storage capacity of liver is limited. Bone Mn is not readily available
- Coloured hair contains more than white
- Hair manganese not related to dietary Mn. Over 50% hair Mn can be derived from external deposition from the environment
- Newborn calf thought to have different Mn metabolism to older animals

Excretion
- Well controlled and mainly lost in faeces via bile (95-98%)
- Little is lost in the urine (0.1-0.3%)
- Much is effectively reabsorbed
- Variable absorption and excretion maintain homeostasis
- Body pool is small, so body excretion is often nearly equal to intake

Requirement /Allowances (mg/kg DM)

Rums	NRC	Pigs	NRC	Poultry	NRC	Others	NRC
Calf	40	Creep	4	Chick	60-30	Dog	5
Dairy	12-24	Weaner	4	Broiler	60	Cat	7.5
Beef	20-40	Grower	3	Breeder	20	Horse	40
Lamb	40	Finisher	2	Layer	20	Fish	2.4-13
Sheep	20-40	Sow/Boar	20	Turkey	60	Rabbits	2.5-8.5

Rums	Typical	Pigs	Typical	Poultry	Typical	Others	Typical
Calf	40-50	Creep	50	Chick	70-80	Dog	30
Dairy	40-100	Weaner	50	Broiler	100	Cat	10
Beef	40-100	Grower	40	Breeder	100	Horse	40-100
Lamb	40	Finisher	40	Layer	40-80	Fish	
Sheep	50	Sow/Boar	40	Turkey	80-120	Rabbits	50

Notes
EU regulation maximum: 150mg/kg in feedstuff for producing animals.

Adequate Status

Species	Liver	Kidney	Blood	Milk	Bone Ash
Cattle	2.5-6	1.2-2.0	0.07-0.09	20-70	
Sheep	2-4.4	0.8-2.5	0.02-0.025		3-6.5
Pigs	2.3-4	1.3-2	0.04	200	
Poultry	2-4		0.085-0.091		7-13
Horses	1-6	0.5-2.4			
Rabbits	1-2	2-3			
Dogs	3-5	1.2-1.8	0.02		
Units	ppm WW	ppm WW	ppm WW	mg/100ml	ppm WW

- Most sensitive non-invasive evaluations are serum, urinary and lymphocyte Mn and SOD activity. Liver Mn not entirely reliable unless severe deficiency
- Egg yolk Mn is 4-5 times that of egg white (egg yolk -33ppm ww)
- Monitoring feed intake appears to be one of best means of assessing Mn status

CONTEXT www.contextbookshop.com

Manganese

15c

| Mn | Atomic Wt 54.94 | Atomic No 25 | Micro Mineral |

Deficiency

General
- Impaired fat and carbohydrate metabolism
- Affects membrane integrity
- Poor growth
- Skeletal abnormalities
- Shortening and bowing of joints
- Reproductive failure (impaired or irregular oestrus, foetal resorption, foetal deformities, non-viable young)
- Nervous disorders especially in young stock
- Reduced Vitamin K-induced clotting response

Ruminants
- Bone malformation in newborn/young, (knuckles bent back)
- Low fertility in adults
- Ataxia (nervous disorders)
- Muscular weakness
- Reduced milk production and level Mn in milk
- Impaired glucose tolerance

Pigs
- Lameness, crooked, shortened legs and enlarged hocks
- Poor mammary development and lactation

Poultry (Chicks)
- Chondroitin sulphate content of epiphyseal cartilage is markedly reduced
- Perosis (but may also appear if choline or niacin are deficient.)
- Nervous signs chacterised by head retraction, 'star gazing'
- Chondrodystrophy from Mn deficient diet of breeding hen. (defective growth, oedema, bone disease, high mortality)
- Poor hatchability

Poultry (Adults)
- Reduced egg production and shell quality
- Hock disorders in turkey poults
- Pancreatic abnormalities

Mammals
- Delayed oestrus, smaller litter size, poor spermatogenesis

Rabbits
- Crooked front legs
- Testicular degeneration

Horses
- Not reported under field conditions

Toxicity

General
- Least toxic of trace minerals
- Poor growth
- Depressed appetite and feed intake
- Leg stiffness
- Anaemia
- Abominal discomfort
- Abortion and cystic ovaries may be associated
- Excess dietary Mn leads to accumulation in thyroid (in rats) which may induce I deficiency

Maximum Dietary Tolerable Level (ppm DM)

Ruminants	1000	Rabbits	400
Pigs	400	Horses	400
Poultry	2-3000		

Interrelationships
Calcium, Phosphorous, Iron, Cobalt, Cadmium, Zinc.

Has a direct effect on:
- The function of copper and zinc and iron may be interchangeable in some enzyme systems
- A manganese/choline interaction maybe significant in 'fat cow syndrome'

Manganese

15d

| Mn | Atomic Wt 54.94 | Atomic No 25 | Micro Mineral |

Antagonists

- Heavy liming, soil pH >6 decreases availability of manganese
- Excess calcium reduces absorption of Mn in duodenum
- Excess Mn interferes with calcium, phosphorus and can induce iron-deficiency anaemia due to its effect on haemoglobin synthesis
- High intakes of Ca and P aggravate Mn deficiency e.g. perosis
- Mn competes with Fe and Co for binding sites on intestinal uptake proteins
- Status can be affected by the amount of protein present
- Mn antagonises molybdenum and dietary Zn reduces Mn absorption
- Excess NDF and phytate can reduce absorption
- Lead toxicity reduces kidney Mn

Synergy

- Mn is necessary for utilisation of biotin, vitamin B_1 and Vitamin C
- Manganese and Vitamin K work together in blood clotting
- Oestrogenic hormones increase absorption
- Low dietary Mn reduces accumulation of Se in tissues
- Mg can effectively substitute for Mn (when in deficiency) in some enzymes

Main Supplements

Source	Element %	Relative Bio-avail.	Comments
Manganese carbonate	43	low	
Manganese chloride	27.5		
Manganese citrate			
Manganese dioxide	36	low	
Manganese gluconate			
Manganese methionine	15	high	
Manganese orthophosphate			
Manganese oxide	52-62	medium	Max. 100ppm Pb or As. <10% as dioxides Low water solubility High hygroscopicity
Manganese phosphate			
Manganese proteinate	10	high	
Manganese sulphate	28.5	high	Water soluble High hygroscopicity

Manganese

15e

| Mn | Atomic Wt 54.94 | Atomic No 25 | Micro Mineral |

Feed Name	mg/kg DM
Alfalfa meal	30.0
Bakery waste	71.0
Barley grain	17.4
Bean field	17.2
Blood meal	6.0
Brewers grains	35.6
Buckwheat grain	38.0
Buttermilk dehyd. (cattle)	4.0
Casein dehyd. (cattle)	5.0
Cassava tubers dehy	30.0
Citrus pulp dried	7.0
Copra meal	71.0
Cottonseed Whole	11.0
Cottonseed meal	25.0
Distillers grains - wheat	49.0
Distillers grains maize	35.6
Distillers grains - barley	24.0
Fishmeal (Sth Am)	21.9
Grass bluegrass	55.0
Grass alfalfa	43.0
Grass bermuda	-
Grass clover	123.0
Grass extensive	55.0
Grass kikuyu	40.0
Grass timothy	144.0
Groundnut ext	55.0
Hay alfalfa	31.0
Hay bluegrass	70.0
Hay clover	73.0
Hay eragostus	74.0
Hominy feed	14.5
Linseed meal (mech ext)	41.7
Maize bran	17.8
Maize germ ext (sol)	8.5
Maize gluten 20	4.4
Maize gluten 60	25.6
Maize grain	7.0
Malt culms	35.5
Milk (cattle-dehyd)	0.5
Milk skimmed	1.6
Millet grain	32.0
Molasses -beet	30.0
Molasses - cane	54.0
Oat groats	31.4
Oat middlings	48.4
Oatfeed	-
Oats grain	48.0
Palm kernel exp	252.7
Peas	13.5
Potato dried	7.5
Rape ext (mech)	61.1
Rice bran	253.0
Rice grain	20.0
Rye grain	66.0
Safflower ext. solv.	20.0
Sesame ext mech	52.0
Silage alfalfa	40.0
Silage grass	75.0
Silage maize	21.0
Silage sorghum	73.0
Silage wholecrop	36.0
Sorghum grain	18.0
Soya ext. solv	39.0
Soya flour	32.0
Soya hipro	39.8
Straw barley	35.0
Straw oat	38.0
Straw wheat	38.0
Sugar beet pulp (dehyd)	38.0
Sugar beet pulp (mol)	28.1
Sunflower ext	48.6
Triticale grain	31.6
Wheat (caustic)	34.9
Wheat bran	152.0
Wheat feed	78.7
Wheat germ ext.	169.0
Wheat grain	34.9
Whey low lactose	9.0
Whey (cattle dehyd)	6.0
Yeast (brewers dehyd)	6.0
Yeast (torula dehyd)	9.0

CONTEXT

www.contextbookshop.com

Bioplex® Manganese from Alltech

| Mn | Atomic Wt 54.94 | Atomic No 25 | Micro Mineral |

Introduction
- Bioplex® Manganese is an organic trace mineral proteinate for use in livestock feeds.

Concentrations Available

Product	Guaranteed Analysis	Ingredients
Bioplex® Mn 10%*	Min. 10% Manganese	Manganese proteinate
Bioplex® Mn 15%*	Min. 15% Manganese	Manganese proteinate

*Not all mineral concentrations are available in every country. Contact your local Alltech representative for details.

Physical Characteristics

Appearance
Bioplex® Manganese is a beige coloured, free flowing powder with no discernible odour.

Storage
Bioplex® Manganese should be stored in a cool, dark area. Shelf life under these conditions is 36 months.

Packaging
Bioplex® Manganese is available in 25 kg bags.

Inclusion Rates of Bioplex® Minerals

- Every species requires a different inclusion rate. Also inclusion depends on the motivation for organic supplementation. There are two issues to be examined.
- Organic supplementation can be for performance reasons, here research shows that supplementation with Bioplex® minerals at rates between 20-50% of inorganic minerals provides superior performance effects.
- Supplementation may be driven by the requirement to meet environmental regulations. Bioplex® Manganese can be used as a complete replacement of inorganic salts at a significantly lower inclusion level.

For inclusion recommendations for your region and other details, contact your local Alltech representative.

Benefits (Conditions responsive to improved Manganese status)
- Manganese is necessary for enzyme activity
- Involved in lipid protein and carbohydrate metabolism
- Proper functioning of reproductive processes
- Growth rate and feed efficiency
- Hormone production
- Bone development and cartilage formation
- Immune function

Some Symptoms of Manganese Deficiency
- Infertility
- Reproductive failures
- Nerve problems
- Poor muscle co-ordination
- Retarded growth

Alltech
...naturally

www.alltech.com

Mercury

Hg | Atomic Wt 200.59 | Atomic No 80 | **Highly Toxic Mineral**

Introduction
- A toxic element from both ingestion and inhalation
- Widely distributed in air, soil, rocks, water, either directly or indirectly, as a result of human activity
- Main industrial source is from chloralkali industry
- Also used in manufacture of electrical apparatus, paint, dental preparations, pharmaceuticals, paper and pulp industry
- Inorganic Hg preparations used as wound dressings, counter irritants and vesicants
- Used in agriculture for seed treatment as methlymercury (v. hazardous)

Key Natural Sources
- Pastures and crops contain little Hg (<0.1ppm DM)
- Feed or water contamination
- Fishmeal would contain highest available Hg
- Consumption of predatory fish from contaminated waters

Absorption
- Readily absorbed by respiratory, gastrointestinal tracts and through unbroken skin
- Ingested inorganic Hg absorption estimated 15%
- Methyl mercury is 60-100% absorbed by ruminants, aquatics, fowls and humans
- Young mammals absorb even inorganic Hg efficiently- (30-40%)

Metabolism
- Little organic or inorganic mercury passes to milk
- Blood, urine, faecal and milk levels have little relationship to Hg toxicity until tissues have reached saturation and damage is excessive
- Egg white accumulates approx. 3x Hg of yolk
- White meat accumulates approx. 2xHg of dark meat
- All forms cross the placenta to the foetus

Excretion
- Only slowly excreted via faeces and urine
- Acts as a cumulative poison

Adequate Status

Species	Liver	Kidney	Blood	Milk	Hair
Cattle	0.0007-0.06	0.008-0.09	<0.1	3-10	0.1
Sheep	<0.01-0.1	0.01-0.04	<0.1		
Pigs	<0.01-0.03	<0.01-0.09			
Poultry	0.01-0.1	<0.02-0.3	0.1		
Horses	<0.1	<0.1	<0.01-0.1		
Cats	0.01-0.1	<0.01-0.1	0.1-0.3		
Units	ppm WW	ppm WW	ppm WW	ppb DM	ppm DM

Notes
Levels in feathers, hair and erthyrocytes are useful indicators of tissue status

Mercury

16b

Hg | Atomic Wt 200.59 | Atomic No 80 | **Highly Toxic Mineral**

Toxicity

General
- Not common and usually related to animals which have ingested contaminated feed or water
- Primarily neurological signs: tremors, vertigo, irritability, moodiness, depression
- Salivation, diarrhoea
- Loss of vision and hearing
- Renal failure
- Chronic exposure: affects reproductive, renal and cardiovascular functions

Ruminants
- Bloody diarrhoea, excessive thirst, salivation, depression, unsteady gait

Pigs
- Anorexia, inability to drink, blindness, ataxia, hypersensitivity, recumbency, death

Poultry
- Loss of appetite, wing and muscle weakness, impaired testicular development, depressed fertility of eggs

Maximum Dietary Tolerable Level (ppm DM)			
Ruminants	2	Rabbits	2
Pigs	2	Horses	2
Poultry	2		

Notes
Based on human food residue considerations

Antagonists
- Methyl Hg intoxication reduces GSH-Px activity (Se based antioxidant enzyme)

Synergy
- Dietary Se compounds protect against Hg toxicity

Main Supplements
- Supplementation not recommended / required
- Vitamin E also has a protective effect against methyl Hg

Molybdenum

17a

Mo | Atomic Wt 95.94 | Atomic No 42 | Micro Mineral

Introduction
- A metallic, silvery-white element, Present in nature in the form of molybdenum sulphide (molybdenite) and lead molybdate
- Abundance in earths crust approximately 1.5mg/kg
- Discovered in 1778 by swedish scientist Carl Wilhelm Scheele
- Name 'Molybdenum' comes from Greek word '*molybdos*', which means lead
- The two largest uses are as an alloy in stainless steels and in alloy steels
- General uses for molybdenum compounds include machinery, electrical applications, in chemicals, pigments
- Used in agricultural industry for direct seed treatment or fertiliser formulation
- An essential trace mineral for plants and animals
- Soil that has no Mo cannot support plant life
- Found in plants from <0.5 to 100ppm DM
- Found in the body at approximately 1-4 mg/kg body weight
- Important as copper antagonist but also prevalent as toxic element for grazing ruminants
- Monogastric diets based on grains are generally low in Mo

Key Natural Sources
- Present at low levels in a wide range of plant and animal tissues
- Concentrations in normal herbage may range from 0.1 to 0.3 ppm DM
- Poorly drained soils, peat, organic, alkaline soils and liming increases Mo uptake by plants reduces sulphate uptake. Legumes generally accumulate more Mo than grasses as it helps nitrogen fixation
- Many plants grown on deficient soils develop deficiency disease
- Mo in plants is present as soluble ammonium molybdate, insoluble molybdenum trioxide, calcium molybdate and molybdenum disulphide
- High industrial activity can increase herbage levels (100-200ppm)
- Highest levels found in shellfish
- Good sources of Mo are sorghum, leafy vegetables, legumes, grains, organ meats, milk, eggs. Poor sources: fruits, root vegetables, muscle meat
- 40% of cereal Mo can be lost in milling

Function
- Component of 6 enzymes, three in mammals
- Xanthine oxidase is essential in the formation of uric acid. Birds excrete excess nitrogen in the form of uric acid
- Aldehyde oxidase may be involved in niacin metabolism
- Sulphite oxidase converts sulphite (derived from S-amino acids, methionine and cysteine) to sulphate for final excretion in urine
- All Mo-metalloenzymes are also Fe-metalloenzymes
- A component in the enamel on teeth
- In plants, stimulates N fixation and reduction. Mo is an electron carrier in enzymes that catalyse the reduction of nitrogen and nitrate

Benefits
- Plays a role in protein synthesis and oxidation reactions
- Helps in the metabolism of fats and carbohydrates
- Necessary for the production of uric acid and taurine
- Involved in DNA metabolism

Absorption
- Readily and rapidly absorbed from most diets
- Absorbed as molybdate from the small intestine with some in the lower intestine
- Extent of absorption depends on species, age of animal, level and chemical nature of Mo in diet
- Average absorption 20-30%
- Water soluble molybdates and Mo in herbage – 75-97% absorbed
- Insoluble molybdates (disulphides) poor absorption
- Rate of absorption and retention is inversely related to level of dietary S
- Sulphate thought to inhibit membrane transport of molybdate, decreasing absorption in intestine and reabsorption by renal tubules

CONTEXT www.contextbookshop.com

Molybdenum

Mo | Atomic Wt 95.94 | Atomic No 42 | Micro Mineral

Metabolism
- Blood Mo loosely attached to red blood cells
- Most transported to liver and converted to molybdate.
- Little storage, mainly in the liver and bones
- Low concentration found in all tissues and fluids

Excretion
- Rapidly excreted in the urine
- Can be reabsorbed by kidney to liver, lungs, bone, skin
- Small quantities lost through bile and faeces (level higher in ruminants due to limited absorption)
- Milk levels vary with dietary Mo intakes
- Mo crosses the placenta

Requirement /Allowances
- No precise requirements
- Animals perform normally on extremely low dietary Mo levels
- Goats estimated requirement 0.05-0.1 ppm DM
  Chicks <0.2 ppm DM
- Requirements are affected by the sulphate level in the diet
- Ruminant levels difficult to estimate due to Cu/S interaction

Adequate Status

Species	Liver	Kidney	Plasma	Milk	Diet
Cattle	0.14-1.4	0.22-0.57	0.01-0.05	18-120	0.5-3.5
Sheep		1.5-6.0			0.5-3.5
Goats					0.05-0.1
Pigs Sows	1.5-6.0		0.01-0.1		0.5-2.5
Poultry					0.03-1.0
Horses					0.5-2.5
Units	ppm WW	ppm WW	ppm WW	ppb DM	ppm WW

Deficiency

General
- Deficiency unlikely to be directly attributable to Mo, naturally
- Reduced feed intake and growth
- Impaired reproduction, infertility
- Elevated mortality in mothers and offspring
- Most often related to copper and sulphur excess

Humans
- Increased incidence of dental carries

Ruminants
- Loss of weight, emaciation
- Reduced conception, increased abortion rate
- Diarrhoea
- Decreased efficiency of cellulose digestion
- May lead to increased incidence of copper toxicity

Pigs
- Increased rate of anoestrus in sows

Poultry Chicks
- High incidence of late embryonic mortality
- Mandibular distortion, anophthalima
  Poor hatchability, weak chicks
- Defects in leg bone development and feathering

Soil
- Reduces plant Vitamin C production and increases nitrate accumulation, particularly in root crops
- Resultant Vitamin C deficiency reduces ability of animals to metabolise and excrete nitrite

Molybdenum

| Mo | Atomic Wt 95.94 | Atomic No 42 | Micro Mineral |

Toxicity

General
- Tolerance to high dietary Mo intake varies with species, age of animal, quantity, chemical form, Cu status of animal and diet, forms and concentration of dietary S
- Tolerance varies with intake of methionine, cystine, and protein which are capable of being oxidised to sulphate in the body
- Typically >5mg/kg bodyweight toxic. Liver levels increase but return to normal when excess intake ceases. If the excess intake is prolonged, a depletion of tissue copper levels occurs and a copper deficiency results
- Copper deficiency may occur when forages have copper levels below 5 ppm, and molybdenum levels above 3 to 5 ppm

Ruminants
- Delay puberty onset
- Decreases conception rate, causes anovulation and anoestrus. (decreased LH release)
- Testicular degeneration
- Increases nitrates in the rumen
- Reduces crude protein entering small intestine
- Intense liquid diarrhoea
- Inappetance, loss of bodyweight, poor growth
- Harsh, staring discoloured coats,
- Weakness, stiffness
- Affects P metabolism
- Affects tissues and bone fractures
- Anaemia (disturbs Fe metabolism)
- Dietary levels 5-6ppm DM inhibit Cu storage
- Symptoms mainly those of Cu deficiency
- Toxicity reported at 100-200mg/kg DM in diet

Pigs
- Most tolerant

Poultry
- May reduce egg production, chick embryonic survival and growth rates, above 350 mg/kg; egg hatchability above 500ppm

Horses
- Seem fairly resistant to molybdenosis. 5-25ppm Mo in forage has shown some disturbance in copper utilisation

Maximum Dietary Tolerable Level (ppm DM)

Ruminants	10	Rabbits	500
Pigs	20	Horses	5
Poultry	100		

Interrelationships
- Availability affected by Cu, Zn, Mn, Cd, S, Tungstate(W), Fe, methionine, cystine
- Tissue level influenced by Fe, Zn, Pb, W, ascorbic acid and a-tocopherol. Copper-molybdenum-sulphur interaction important

Antagonists
- Excess copper and/or sulphate interferes with molybdenum utilisation
- Mo reduces Cu deposition in organs and the synthesis of ceruloplasmin. High dietary Cu reduces Mo deposition in liver
- Level of sulphate in the diet can alter the absorption, retention, and excretion of absorbed molybdenum
- High sulphate increases urinary Mo and decreases tissue deposition
- Tungstate (W) competitively inhibits utilisation of Mo for formation of xanthinie dehydrogenases. W increases Mo urinary excretion Silicates inhibit absorption of dietary molybdate
- High sulphates intake can help counteract the effect of high molybdenum levels in forage
- Some indication that in ruminants, low levels of cobalt intake and increased molybdenum may suppress vitamin B_{12} synthesis

Molybdenum

17d

Mo | Atomic Wt 95.94 | Atomic No 42 | Micro Mineral

Synergy
- Can help control chronic copper poisoning

Main Supplements
- Does not usually need to be added to practical diets
- Some areas Mo may be added to sheep rations to prevent copper toxicity. This is under veterinary direction
- Typical products that could be used in supplements

Main Supplements (Typical products that could be used in supplement)

Source	Element %	Relative Bio-avail.	Comments
Sodium molybdate	38	high	water soluble skin irritant
Ammonium molybdate	54	high	water soluble
Magnesium molybdate		low	insoluble in water
Calcium molybdate		low	insoluble in water

CONTEXT www.contextbookshop.com

Molybdenum

17e

| Mo | Atomic Wt 95.94 | Atomic No 42 | Micro Mineral |

Feed Name	mg/kg DM
Alfalfa meal	0.7
Bakery waste	-
Barley grain	0.4
Bean field	0.6
Blood meal	0.2
Brewers grains	0.6
Buckwheat grain	5.5
Buttermilk dehyd.(cattle)	-
Casein dehyd. (cattle)	-
Cassava tubers dehy	0.1
Citrus pulp dried	0.4
Copra meal	0.4
Cottonseed whole	-
Cottonseed meal	2.5
Distillers grains - wheat	0.6
Distillers grains maize	1.5
Distillers grains- barley	-
Fishmeal (Sth Am)	0.4
Grass bluegrass	-
Grass alfalfa	-
Grass bermuda	-
Grass clover	-
Grass extensive	3.0
Grass kikuyu	1.3
Grass timothy	-
Groundnut ext	-
Hay alfalfa	0.7
Hay bluegrass	-
Hay clover	-
Hay eragostus	-
Hominy feed	0.2
Linseed meal (mech ext)	0.7
Maize bran	-
Maize germ ext (sol)	-
Maize gluten 20	0.3
Maize gluten 60	1.3
Maize grain	0.6
Malt culms	0.6
Milk (cattle-dehyd)	0.3
Milk skimmed	0.3
Millet grain	0.2
Molasses -beet	0.3

Feed Name	mg/kg DM
Molasses - cane	-
Oat groats	-
Oat middlings	-
Oatfeed	-
Oats grain	0.7
Palm kernel exp	0.4
Peas	0.9
Potato dried	0.2
Rape ext (mech)	0.8
Rice bran	0.3
Rice grain	0.6
Rye grain	0.2
Safflower ext. solv.	-
Sesame ext mech	2
Silage alfalfa	-
Silage grass	1.4
Silage maize	0.9
Silage sorghum	-
Silage wholecrop	0.6
Sorghum grain	0.5
Soya ext.solv	3.0
Soya flour	-
Soya hipro	2.8
Straw barley	0.2
Straw oat	0.6
Straw wheat	0.5
Sugar beet pulp (dehyd)	0.4
Sugar beet pulp (mol)	-
Sunflower ext	0.9
Triticale grain	0.5
Wheat (caustic)	0.3
Wheat bran	0.7
Wheat feed	0.8
Wheat germ ext.	-
Wheat grain	0.3
Whey low lactose	-
Whey (cattle dehyd)	-
Yeast (brewers dehyd)	1.0
Yeast (torula dehyd)	0.1

CONTEXT www.contextbookshop.com

Nickel

18a

| Ni | Atomic Wt 58.71 | Atomic No 28 | **Minor Mineral** |

Introduction
- An essential trace element for animals required in very small quantities which are usually present in feed raw materials

Key Natural Sources
- Volcanoes, windblown dust, industrial pollution, coins and Cd batteries
- Found in many rocks as sulphides, oxides, silicates, arsenides and antimonides
- Found in soil usually in clay at about 70mg/kg although the level varies widely
- Feedstuff level is variable and depends on site and species. (usually in the form of an anionic amino complex)
- Alkali soils reduce Ni uptake by plants
- Plant foods greater level than animal origin food
- Rich sources: chocolate, nuts, dried beans, peas, grains
- Pastures typically 0.5-3.5ppm Ni (level decreases with forage maturity)
- Grains: 0.30 to 1.0ppm DM

Function
- Involved in protein and nucleic acid metabolism (RNA and DNA)
- Potential involvement in hormone regulation
- Cofactor or structural component in specific metalloenzymes
- May facilitate intestinal absorption of ferric iron
- May enhance rumen microbial population

Benefits
- Required for maximum activity of rumen bacterial urease. Urease is Ni dependant

Requirement
- Not widely researched as Nickel available in nature exceeds hypothesised requirements
- Ruminants 300-500ppb,
- Monogastric 50-200ppb

Excretion
- The kidneys regulate the excretion

Adequate Status

Species	Liver	Kidney	Serum	Milk
Cattle	0.1-0.6	0.15-0.5	1.2-5.6	0.16-0.8
Sheep	0.05-0.07	0.17-0.4		
Goats			2.7-4.4	<0.12
Pigs	100-200	250-1000	4.2-5.6	0.1
Poultry	0.04-0.1	0.06-0.13		
Horses			1.3-2.5	
Rabbits			6.5-14	
Cats			1.5-6.4	
Dogs	0.04-0.12	0.06-0.12	1.8-4.2ppb	
Units	ppm WW	ppm WW	µg/l DM	ppm DM

Nickel

18b

| Ni | Atomic Wt 58.71 | Atomic No 28 | Minor Mineral |

Deficiency

General
- Not often seen in commercial live stock
- Can be produced in chicks, pigs, goats, rats, lambs, cows
- Poor development of young animals
- Reduced hair growth
- Potentially reduced haemoglobin and erthrocyte levels due to its action on the liver
- Has been reported to reduce zinc, iron and copper levels in the body

Cattle
- Reduces rumen bacterial urease activity

Pigs
- Reduced growth rate and delayed oestrus

Poultry
- Decreased yellow pigmentation of shank skin
- Thickened legs
- Swollen hocks
- Dermatitis of shank skin
- Anaemia

Sheep
- Increased mortality
- Reduced rumen bacterial urease activity
- Reduced liver cholesterol and serum protein (lambs)

Goats
- Increased abortion rate
- Decreased conception rate, viability of young
- Decreased milk production
- Skin lesions
- Lower testicular weights

Toxicity

General
- Relatively non toxic from oral consumption
- Not seen in commercial livestock production unless by error
- The kidneys are the first organ to be damaged by excess levels. >250ppm dietary Ni, toxicosis in rats, mice, chicks, dogs, rabbits, pigs, ducks, monkeys
- Reduced feed palatability/intake
- Reduced growth rates
- Interacts with immune system

Maximum Dietary Tolerable Level (ppm DM)

Ruminants	50	Rabbits	50
Pigs	100	Horses	50
Poultry	300		

Interrelationships
Zinc, Iron, Copper, Manganese
- Iron may need nickel for its absorption and function
- Zinc utilisation is increased in the presence of nickel and it may help in zinc deficiency conditions
- Nickel may replace zinc in some metallo-enzymes
- Effects of Copper deficiency may be temporarily removed by nickel in the short term
- Dietary protein level influences response to Ni supplementation
- Ni deficiency may alter Ca incorporation into the skeleton

Antagonists
- High plant Ni, reduces plant Fe and Mn
- Long term over supply of nickel may reduce copper utilisation
- Bone Calcium declines with Ni deficiency (pigs)
- Requirement increased by pregnancy and high phytate diets (grain and oilmeal)-pigs

Synergy
- Excess cobalt enhances Ni toxicity (poultry)
- Ni-Fe synergy when Fe in the ferrous form

CONTEXT   www.contextbookshop.com

Nickel

18c

Ni | Atomic Wt 58.71 | Atomic No 28 | Minor Mineral

Supplements

Source	Element %	Relative Bio-avail.	Comments
Nickel carbonate hyrdroxide	46		
Nickel chloride	24		
Nickel acetate	23		

Notes
Chloride is more toxic (400ppm) than acetate or carbonate.

Nickel

| Ni | Atomic Wt 58.71 | Atomic No 28 | Minor Mineral |

Feed Name	mg/kg DM	Feed Name	mg/kg DM
Alfalfa meal	-	Molasses - cane	-
Bakery waste	-	Oat groats	0.3
Barley grain	0.3	Oat middlings	-
Bean field	0.1	Oatfeed	-
Blood meal	0.7	Oats grain	0.9
Brewers grains	-	Palm kernel exp	-
Buckwheat grain	1.14	Peas	3.3
Buttermilk dehyd. (cattle)	-	Potato dried	-
Casein dehyd. (cattle)	-	Rape ext (mech)	2.2
Cassava tubers dehy	1.0	Rice bran	-
Citrus pulp dried	-	Rice grain	0.2
Copra meal	-	Rye grain	1.5
Cottonseed whole	-	Safflower ext. solv.	-
Cottonseed meal	-	Sesame ext mech	7.0
Distillers grains - wheat	-	Silage alfalfa	-
Distillers grains maize	-	Silage grass	-
Distillers grains- barley	-	Silage maize	-
Fishmeal (Sth Am)	-	Silage sorghum	-
Grass bluegrass	-	Silage wholecrop	-
Grass alfalfa	-	Sorghum grain	1.1
Grass bermuda	-	Soya ext.solv	21.0
Grass clover	-	Soya flour	-
Grass extensive	-	Soya hipro	-
Grass kikuyu	-	Straw barley	-
Grass timothy	-	Straw Oat	-
Groundnut ext	-	Straw wheat	-
Hay alfalfa	-	Sugar beet pulp (dehyd)	2.2
Hay bluegrass	-	Sugar beet pulp (mol)	-
Hay clover	-	Sunflower ext	-
Hay eragostus	-	Triticale grain	-
Hominy feed	-	Wheat (caustic)	3.4
Linseed meal (mech ext)	-	Wheat bran	1.4
Maize bran	-	Wheat feed	-
Maize germ ext (sol)	-	Wheat germ ext.	-
Maize gluten 20	-	Wheat grain	3.4
Maize gluten 60	5.6	Whey low lactose	-
Maize grain	-	Whey (cattle dehyd)	0.9
Malt culms	0.2	Yeast (brewers dehyd)	5.1
Milk (cattle-dehyd)	0.2	Yeast (torula dehyd)	1.7
Milk skimmed	-		
Millet grain	-		
Molasses - beet	5.7		

CONTEXT www.contextbookshop.com

Phosphorous

19a

| P | Atomic Wt 30.97 | Atomic No 15 | Macro Mineral |

Introduction
- Essential in the formation of bones and body structure as well as key body functions. Makes up about 29% of the total minerals in the body.
- 2nd most abundant mineral in the body 80% occurs in the skeleton, 20% in nucleotides, nucleic acids, phospholipids etc
- Especially concentrated in red blood cells, muscle and nerve tissues. Function is intracellular
- Closely associated with calcium, affecting status of one other. Ca:P ratio in bone is slightly greater than 2:1 and approximately constant
- Does not occur free in nature, as it is much too reactive. Occurs naturally as phosphates

Key Natural Sources
- Found in most common feedstuffs
- Milk a source of high quality P
- Forage levels generally low
- Grains and seed by-products are generally relatively rich
- Levels in feeds are highly variable
- Determined largely by the phosphorous status in the soil, the stage of maturity and the plant, crop management, climate and soil pH
- P concentration declines as forage matures

Phytate
- Most of phosphorous in cereals, oilseeds and co-products is in the organically bound phytin form, (50-70%) which will be shown as P on chemical analysis
- Phytate P can be utilised by ruminants(80%) but not by non-ruminants i.e. Pigs (30-40%) and chickens (25-45%). Turkeys are thought to be able to use it more effectively at closer to 80%. Horses some utilisation in hindgut
- Non-ruminants are dependent on the presence of phytase in the feedstuffs or intestinal secretions to utilise phytate phosphorous
- Older animals have a greater ability to digest phytate than younger animals
- Pelleting has been shown to reduce the levels of phytase naturally present, which would release phosphorous. Additional phytase will also improve the absorption of calcium in pigs and poultry possibly due to the increase calcium need as more phosphorous is made available
- Added calcium may reduce the efficiency of phytase
- Commercial phytase added to rations in monogastrics, increases bio-availability of P and so reduces excretion of P

Function
- Involved in almost all metabolic reactions
- Acts as a cofactor for enzymes and activates the B-complex vitamins
- Involved in bone mineralisation
- Amino acid, carbohydrate and fat metabolism
- Enzyme formation
- Component of nucleic acids (RNA and DNA)
- Involved in cellular communication
- Involved in sugar digestion and energy production
- Maintenance of osmotic and acid base balance
- Phospholipid formation
- Protein formation (e.g.. part of casein)
- Required by rumen microorganisms for cellulose digestion and microbial protein synthesis

Benefits
Involved in
- Energy production
- Milk production
- Bone and tooth formation and maintenance
- Appetite control
- Acid base balance
- Essential for storage, liberation and transfer of energy
- Genetic transmission
- Efficiency of feed utilisation
- Muscle building
- Nucleic acid formation

CONTEXT   www.contextbookshop.com

Phosphorous

| P | Atomic Wt 30.97 | Atomic No 15 | Macro Mineral |

Absorption
- Absorbed throughout the intestinal tract with the duodenum and jejunum being most active
- Ruminants: saliva is a main contributor of P in the gut
- Absorption of phosphorus is related to the supply of available absorbable phosphorous in the intestine
- Intestinal absorption efficiency may vary between 50-70% and even lower for horses 30-50%
- High% of P in plants is in the phytate form, which is absorbed with differing ability depending on the animal. (see above)

Absorption and so requirements affected by
- Calcium to Phosphorous ratio
- DM and Phosphorous intake
- Variability of ingredient nutrients
- Physiological status /Production requirements (milk, growth, pregnancy)
- Breed
- Activity level
- Phosphorous source / absorption capacity
- Intestinal pH
- Vitamin D intake and adequacy of liver and kidney
- Age (young animals higher absorption)
- Intestine parasitism
- Environmental factors including stress from disease, overcrowding, poor ventilation and inadequate temperature control
- Dietary intake of calcium, iron, aluminium, magnesium, manganese, potassium
- Increased by Vitamin C
- Decreased by phytase

Metabolism
- After absorption, phosphorus is stored, secreted into the GIT lumen for reabsorption, or used for bone formation, milk or eggs etc.
- Blood level is controlled by the parathyroid hormone and thyrocalcitonin
- It can be withdrawn from the bones to maintain normal plasma levels
- Mobilisation of calcium from the bone releases phosphate ions into the blood circulation

Excretion
- Herbivores: Faeces is primary path unless on high concentrate diet
- Carnivores and humans: Urine is primary route
- Ruminants: can also excrete excess absorbed phosphorus via salivary secretion
- Variable faecal excretion is an important homeostatic control
- Some excreted P is in the form of microbial cells and their products
- In short supply the kidney tubules conserve phosphorous by returning it to the blood and when excess, they excrete more to the urine
- Low availability phytate P increases excretion of P in manure; (environmental impact)

Phosphorous

| P | Atomic Wt 30.97 | Atomic No 15 | Macro Mineral |

Requirement /Allowances (g/kg DM)

Rums	NRC	Pigs	NRC	Poultry	NRC	Others	NRC
Calf (a)	2.8	Creep	7	Chick	4.5	Dog	9-16
Dairy	3.5-4.5	Weaner	6.5	Broiler	3.5	Cat	6
Beef	0.5-2.5	Grower	6	Breeder	2.5	Horse	1.7-3.8
Lamb		Finisher	4.5	Layer	2.5	Fish	4.5-6
Sheep	1.6-3.8	Sow/Boar	6	Turkey	3.5-6	Rabbits	2.2-5.0

Rums	Typical	Pigs	Typical*	Poultry	Typical*	Others	Typical
Calf	4	Creep	4.5	Chick	4-4.5	Dog	5-10
Dairy	3.5-4.5	Weaner	4.5	Broiler	4	Cat	5-8
Beef	1.5-2	Grower	3.5	Breeder	4-4.3	Horse	1.5-5
Heifer	2.5-3.1	Finisher	3.0	Layer	4-5	Fish	
Sheep	2.5-5	Sow/Boar	4.0	Turkey	5-7	Rabbits	

<u>Notes</u>
(a) Milk replacer 4.5-7 g/kg DM
(b) Milk replacer 6-7 g/kg DM
  * Available Phosphorous

Adequate Status

Species	Liver	Kidney	Serum	Urine	Milk
Cattle	6-14	8-13	4.5-6.0		600-1000
Sheep	7-14	6-13	4-8	0.1-1.5	
Pigs	12-14	10-12	6-10.7	0.02-0.8	1400-1700
Poultry			4.5-6.0		
Horses	8-12	9-11	2.7-5.0	0.25-2.3	50-90
Dogs	6-12	7-14	3.5-6.0		
Units	mg/g DM	mg/g DM	mg/100ml DM	mg/100ml DM	mg/l DM

- Pigmented/coloured hair contains more P than white hair. Levels do not correlate with dietary intake
- Typical levels for eggshells are >0.12%

Phosphorous

19d

| P | Atomic Wt 30.97 | Atomic No 15 | Macro Mineral |

Deficiency

General
- Weakness
- Demineralisation of bone (similar effects as Ca deficiency)
- Loss of calcium
- Poor appetite (anorexia may develop in extreme cases)
- Disturbance of energy metabolism
- Poor growth
- Poor feed conversion
- Pica (depraved appetite)
- Dull, dry hair coats, listless
- Poor fertility (failure to show oestrus, low conception rate)
- Reproductive failure
- Severe deficiency leads to rickets (young), osteomalacia (older) and other related diseases

Ruminants
- Grazing livestock, P most prevalent mineral in deficiency
- Infertility: irregular oestrus, silent heat, delayed/low conception
- Reduced milk yield and quality. Initially able to draw on skeletal reserves to maintain yield
- Milk fever (Mg, Ca link)
- Reduced milk protein potential
- Chewing of wood, rocks, bone
- Low ruminal P reduces fibre digestion, microbial protein synthesis
- Poor lambing rates
- May reduce metritis and reduce immune response

Poultry
- Reduced egg production, quality, size, reduced shell thickness
- Impaired reproduction (poor hatchability, dead, weak or formed offspring, decreased mating activity, delayed sexual maturity)
- Increased mortality
- Increased thyroid size

Pigs
- Bone structure abnormalities
- Lameness
- Stilted gait
- Posterior paralysis
- Fractures
- Bone softness
- Unthrifty
- Beading of ribs
- Enlargement and erosion of joints

Horses
- Craving for and chewing bones, wood, hair, rocks, clothing

Toxicity

General
- Not considered toxic form single large dose or consumption
- Mild diarrhoea may occur
- Prolonged high intakes
- Reduce fertility
- Impair skeletal growth
- Impair manganese availability
- Increase need for Ca, Fe, Mg, Mn
- Increase urinary calculi risk
- Excessive bone resorption

Ruminants
- Excess P in relation to Ca, 'Downer cow'
- Myopathy and skeletal softening, weak bones
- Higher risk of urinary calculi (more in sheep than cattle)
- Reduce apparent absorption of magnesium
- Inhibit Vitamin D activity

Horses
- Hyperparathyroidism (NSH) where excess phosphorous is fed in diets of low calcium. Is seen as lameness and enlargement of lower jaws etc. "big head"
- Leads to calcium deficiency

Poultry
- Excess acidic sources such as monobasic phosphate affect the acid-base balance on laying hens
- Restlessness, irritability, cannibalism
- Reduced egg production and egg shell quality
- Increase leg deformities, mortality

Maximum Dietary Tolerable Level (g/kg DM)

Cattle	10	Rabbits	10
Sheep	10	Dogs	16
Pigs	15	Horses	10
Poultry	10 (8 for layers)	(Ratio of Ca to P is important)	

CONTEXT   www.contextbookshop.com

Phosphorous

| P | Atomic Wt 30.97 | Atomic No 15 | Macro Mineral |

Interrelationships
- Metabolism and requirements affected by Ca, Na, Mg, Fe, Zn, Mo, Mn, Cu, Al, Vit D

Ratio of Ca:P
- Can be important and is normally between 1:1 and 2:1, but the level is more important. If ratio is less than 1:1, then calcium absorption will be affected, and the excess phosphorous will cause bone malformations
- The effect of high calcium levels depends on the phosphorous level and whether it is adequate. Soluble phosphorous in the gut is reduced when the level of dietary phosphorous rises above 150% of the calcium requirement

Pigs	1:1 to 2:1 (diet dependant)
Horses	1:1 to 2:1
Cattle	>1:1 to maximum 7:1
Sheep	2:1 max (4:1)
Chickens	2:1 (layers: increase Ca ratio)
Dogs	1.3:1
Pigs max.	1.3:1 in low P diets, 2:1 in high P diets.

Antagonists
- Phytate Phosphorous is poorly utilised by non-ruminants
- Phytin (a complex of inositol, P and other minerals) from plants reduces availability
- High dietary P may reduce Se retention (pigs)
- Monogastrics- high P intensifies effect of Mn deficiency
- High Mo may disturb P metabolism (ruminants)

Synergy
- Ruminants: providing they have a good Vitamin D status can tolerate a wide range of dietary levels of Ca: P

Main Supplements
- The choice of supplementary mineral source depends on the chemical composition, biological availability, cost, impurities present and even the dust hazard present

Main Supplements			
Source	Element %	Relative Bio-avail.	Comments
Sodium phosphate	21-25	high	Corrosive (water soluble)
Ammonium polyphosphate			Corrosive (water soluble)
Phosphoric acid	23-55	high	
Potassium phosphate	22.8	high	
Deflourinated rock phosphate	13.3		(Range is 8.7-21 %)
Bone meal steamed	8-18	high	
Dicalcium phosphate	18.5	high	
Monocalcium phosphate	18.6-21	high	
Soft rock phosphate	9	low	Fluorine
Trical phosphate	18		
Mono-ammonium phosphate			
Diammonium phosphate			
Sodium tropolyphosphate			

Phosphorous

19f

| P | Atomic Wt 30.97 | Atomic No 15 | Macro Mineral |

Feed Name	g/kg DM	Feed Name	g/kg DM
Alfalfa meal	2.2	Molasses - cane	1.1
Bakery waste	2.6	Oat groats	4.6
Barley grain	4.1	Oat middlings	4.8
Bean field	5.6	Oatfeed	1.8
Blood meal	2.6	Oats grain	3.6
Brewers grains	4.4	Palm kernel exp	5.5
Buckwheat grain	3.7	Peas	4.4
Buttermilk dehyd. (cattle)	10.1	Potato dried	2.3
Casein dehyd. (cattle)	9.0	Rape ext (mech)	11.7
Cassava tubers dehy	1.3	Rice bran	17.1
Citrus pulp dried	1.1	Rice grain	3.2
Copra meal	6.6	Rye grain	3.7
Cottonseed whole	7.5	Safflower ext. solv.	8.1
Cottonseed meal	9.0	Sesame ext mech	14.6
Distillers grains - wheat	8.7	Silage alfalfa	3.0
Distillers grains maize	5.4	Silage grass	2.8
Distillers grains - barley	9.7	Silage maize	1.8
Fishmeal (Sth Am)	27.3	Silage sorghum	2.1
Grass bluegrass	3.4	Silage wholecrop	1.8
Grass alfalfa	3.0	Sorghum grain	3.3
Grass bermuda	2.1	Soya ext. solv	7.3
Grass clover	3.5	Soya flour	1.8
Grass extensive	2.8	Soya hipro	6.8
Grass kikuyu	4.3	Straw barley	0.7
Grass timothy	2.8	Straw oat	0.1
Groundnut ext	5.5	Straw wheat	0.7
Hay alfalfa	2.8	Sugar beet pulp (dehyd)	1.0
Hay bluegrass	2.5	Sugar beet pulp (mol)	0.79
Hay clover	2.5	Sunflower ext	5.6
Hay eragostus	4.4	Triticale grain	3.3
Hominy feed	5.1	Wheat (caustic)	3.8
Linseed meal (mech ext)	9.1	Wheat bran	13.6
Maize bran	3.9	Wheat feed	9.0
Maize germ ext (sol)	5.6	Wheat germ ext.	10.5
Maize gluten 20	8.9	Wheat grain	3.8
Maize gluten 60	4.5	Whey low lactose	11.2
Maize grain	2.9	Whey (cattle dehyd)	8.2
Malt culms	7.7	Yeast (brewers dehyd)	14.9
Milk (cattle-dehyd)	7.4	Yeast (torula dehyd)	17.1
Milk skimmed	10.6		
Millet grain	3.5		
Molasses -beet	0.3		

Phosphorous Bioavailability

19g

Averages (for swine) relative to availability of P in monosodium phospate at 100% (adapted from Cromwell, 1989)

| P | Atomic Wt 30.97 | Atomic No 15 | Macro Mineral |

Feed Name	%	Feed Name	%
Alfalfa meal	-	Molasses -cane	-
Bakery waste	-	Oat groats	-
Barley grain	31	Oat middlings	-
Bean field	-	Oatfeed	-
Blood meal	-	Oats grain	30
Brewers grains	-	Palm kernel exp	11
Buckwheat grain	-	Peas	-
Buttermilk dehyd.(cattle)	-	Potato dried	-
Casein dehyd. (cattle)	-	Rape ext (mech)	21
Cassava tubers dehy	-	Rice bran	-
Citrus pulp dried	-	Rice grain	-
Copra meal	-	Rye grain	-
Cottonseed Whole	-	Safflower ext. solv.	3
Cottonseed meal	15	Sesame ext mech	-
Distillers grains - wheat	-	Silage alfalfa	-
Distillers grains maize	-	Silage grass	-
Distillers grains- barley	-	Silage maize	-
Fishmeal (Sth Am)	102	Silage sorghum	-
Grass bluegrass	-	Silage wholecrop	-
Grass alfalfa	-	Sorghum grain	19-43
Grass bermuda	-	Soya ext.solv	-
Grass clover	-	Soya flour	-
Grass extensive	-	Soya Hipro	25
Grass kikuyu	-	Straw Barley	-
Grass timothy	-	Straw Oat	-
Groundnut ext	-	Straw wheat	-
Hay alfalfa	-	Sugar beet pulp (dehyd)	-
Hay bluegrass	-	Sugar beet pulp (mol)	-
Hay Clover	-	Sunflower Ext	-
Hay Eragostus	-	Triticale grain	-
Hominy feed	14	Wheat (caustic)	-
Linseed meal (mech ext)	-	Wheat bran	-
Maize bran	-	Wheat feed	35
Maize germ ext(sol)	-	Wheat germ ext.	-
Maize gluten 20	59	Wheat grain	50
Maize gluten 60	-	Whey low lactose	-
Maize grain	14-49	Whey(cattle dehyd)	76
Malt culms	-	Yeast (brewers dehyd)	-
Milk (cattle-dehyd)	-	Yeast (torula dehyd)	-
Milk skimmed	-		
Millet grain	-		
Molasses -beet	-		

CONTEXT www.contextbookshop.com

Potassium

20a

| K | Atomic Wt 39.10 | Atomic No 19 | Macro Mineral |

Introduction
- Discovered by Sir Humphrey Davy in 1807
- Silvery white, light, alkali metal element with brilliant lustre. Burns with violet flame
- Not found free in nature, very strong reducing element
- Name 'Potassium' comes from English word '*potash*' (potassium carbonate) and chemical symbol from latin word '*kalium*'. (may have come from Arabic word, '*gali*', meaning '*alkali*'
- Makes up about 2.6% of earth's crust by weight
- Potash used as a fertiliser, in medicine, in the chemical industry, and is used to produce decorative colour effects on brass, bronze, and nickel
- Essential for plant and animal life
- The third most abundant element in the animal body
- Forms about 0.2-0.3% DM of the body, two-thirds located in skin and muscle
- 98% K found in intracellular fluids

Key Natural Sources
- Forage content affected by : plant maturity, species, variety, management (grazing or crop removal), fertilisation (particularly K and N fertilisers), soil and environmental conditions
- Mature forages and weathered hay exposed to rain and sun are lower in K
- Highest levels in plant leaves compared to seeds
- Good levels found in molasses, potato, soyabean meals, nuts, vegetables, fruit etc
- Very low level feeds tend to be by-product feedstuffs, e.g. brewers grains, maize gluten meal, cotton seed hulls
- Alkali- treated straw (ammonia or sodium hydroxide) reduces K by approx. 25%
- Cereal grain levels tend to be sufficient for monogastrics but not ruminants
- High variability seen in all feedstuffs

Function
- Major cation in intracellular fluid
- Involved in maintaining anion-cation balance
- Nerve impulse conduction
- Muscle contraction
- Fluid transport
- Hormone release
- Embryonic development
- Cellular osmotic balance
- Important for transportation of oxygen and carbon dioxide by the blood
- Cardiac and renal tissue maintenance
- Activator or co-factor of enzymatic reactions, including energy transfer and utlisation, protein synthesis, carbohydrate metabolism
- Water balance

Benefits
- Digestion of food
- Osmotic regulation
- Energy production
- Normal heart activity (relaxes the heart muscle)
- Bacteria have a relatively high requirement for K for normal function and growth
- Cattle in thermo neutral environment, increasing level of dietary K, > 8 to 10 g/kg DM can increase feed intake and milk yield in cows

Absorption
- Absorption mainly by simple diffusion
- Very efficiently absorbed (up to 90-97%) mostly in the duodenum, a little in jejunum, ileum and large intestine
- Ruminants: rumen, omasum and lower GI tract
- Significant amount in rumen is from saliva (high in K)
- While the digestive juices contain high K levels most is reabsorbed
- Low rumen pH impairs transport across the gut wall

CONTEXT   www.contextbookshop.com

Potassium

| K | Atomic Wt 39.10 | Atomic No 19 | Macro Mineral |

Metabolism
- Enters bloodstream largely via conductance channels in membrane of gut mucosa
- K enters cells by an active metabolic process
- K remains unaltered chemically when ingested, retained or excreted
- K concentration in ruminant saliva controlled by aldosterone
- Regulation of K against toxicity better than against deficiency
- Levels increase when there is growth or deposition of lean tissue and are lost when the muscle is broken down at time of starvation, injury or protein deficiency
- Body storage is small, and so should be provided on a daily basis

Excretion
- Mainly via urine by filtration and secretion
- The kidneys regulate the body levels (kidney failure results in excessive levels)
- Adrenal hormones, particularly aldosterone, aids excretion of K by kidney
- Stress tends to increase circulating aldosterone levels, increasing excretion in urine
- Faecal loss is about 13% of total in cows. Lactating cows, 12% of loss is via milk
- Sheep, considerable loss via fleece and skin (particularly hot/humid conditions)
- High urea diets cause high urinary K loss
- High temperatures > 25°C increases excretion

Requirement /Allowances (g/kg DM)

Rums	NRC	Pigs (d)	NRC	Poultry (d)(e)	NRC	Others	NRC
Calf (a)	6.5	Creep	3.0	Chick	2.5	Dog	(b)
Dairy (c)	9-12	Weaner	2.8	Broiler	3.0	Cat	4.0
Beef	6-7	Grower	2.6	Breeder	1.5	Horse	3.0
Heifer	4.6-4.8	Finisher	1.9	Layer	1.5	Fish	
Sheep	5-8	Sow/Boar	2.0	Turkey	4-7	Rabbits*	6.0
Rums	Typical	Pigs (d)	Typical	Poultry (d)(e)	Typical	Others	Typical
Calf	6.5	Creep	3	Chick	4	Dog	4-6
Dairy (c)	9-12	Weaner	3	Broiler	3	Cat	6.0
Beef	6	Grower	3	Breeder	1.8	Horse (f)	3-10
Heifer	5-6.5	Finisher	3	Layer	1.9	Fish	
Sheep	8-20	Sow/Boar	2.5	Turkey	6	Rabbits	

Notes
(a) Milk replacer
(b) 0.09-0.24g/kg BW
(c) Dairy-under conditions of heat stress, increase to 12g/kg DM
(d) Excessive chloride increases requirement for K in pigs and poultry.
(e) Poultry- K requirement increases as dietary protein and temperature increases
(f) Horses- higher levels required for performance horses from sweat loss.

Potassium

20c

| K | Atomic Wt 39.10 | Atomic No 19 | Macro Mineral |

Adequate Status

Species	Liver	Serum	Saliva	Urine	Milk
Cattle		4.0-5.8	50	19-120	150-190
Sheep	0.89-0.93	3.9-5.4			
Pigs		3.5-4.7			
Poultry	0.98-1.3	5.0-6.5			
Horses		2.2			30-65
Rabbits		5.5-6.0			
Dogs		4.2-5.2			
Cats		4.0-5.3			
Goats		4.4-6.7			
Units	% DM	mEq/l DM	mg/100ml DM	mEq/l DM	mg/100ml DM

Notes
- Serum levels should be interpreted with care.
- Dietary status and reduced feed consumption appear to be early signs of deficiency.
- Hair level for cattle 0.2-0.8 % DM
- Saliva k increases with Na deficiency

Deficiency

General
- Inappetance
- Reduced growth
- Dehydration
- Muscle weakness
- Poor intestinal tone
- Cardiac weakness
- Degeneration of vital organs
- Nervous disorders
- Diarrhoea due to stress or disease is common cause of deficiency

Ruminants
- More often seen on high maize silage ration or high dietary urea use (N source with no K)
- Reduced feed and water intake
- Reduced feed efficiency
- Decreased pliability of hide
- Loss of hair condition
- Decreased milk yield and quality
- Lower concentration of potassium in plasma and milk
- Pica: aberrant craving or appetite for unnatural feeds, floor licking, chewing wood etc
- Stress increases requirements and risk of deficiency

Pigs
- Not usual under practical farm conditions
- Rough hair coat, emaciation, inactivity, ataxia

Poultry (Chicks)
- Not usual in commercial units.
- Overall muscle weakness, weak cardiac, respiratory muscles leading to failure and death
- Reduced egg production, egg weight, shell thickness and albumen content
- Mainly occurs during severe stress including heat stress

Horses
- Young horses: reduced growth rate, appetite and hypokalemia
- Pica behaviour

Potassium

| K | Atomic Wt 39.10 | Atomic No 19 | Macro Mineral |

Toxicity

General
- Cardiac insufficiency
- Oedema, muscle weakness, death
- Excess potassium can result from kidney failure
- Heart and adrenal glands adversely affected

Ruminants
- Increase thirst and urination
- Reduced rumen efficiency
- Udder oedema in freshening or milking cows
- Predisposes to hypomagnesemia
- Predisposes to milk fever (effects cation-anion difference)

Pigs
- Hyperkalemia leading to cardiac arrest
- Can tolerate up to 10 times K requirement if plenty of drinking water is provided

Sheep
- Reduced weight gain

Horses
- Muscular disease, hyperkalemic periodic paralysis, a hereditary genetic effect- short episodes of muscle spasms and collapse

Maximum dietary tolerable level (g/kg DM)

Ruminants	30	Rabbits	30
Pigs	20	Horses	30
Poultry	20		

Interrelationships
- An ionic balance exists amongst K, Na, Ca and Mg

Antagonists
- High K interferes with and reduces Mg absorption
- High K increases Ca and Na absorption
- Mg deficiency results in a failure to retain K leading to a deficiency
- Excess salt depletes K, as Na and K must always be in balance
- High temperature (>25°C) and stress increase K requirement

Main Supplements

Source	Element %	Relative Bio-avail.	Comments
Potassium bicarbonate		High	Better palatability
Potassium chloride (a)	50	High	Average palatability
Potassium sulphate	41	High	
Potassium acetate	100		
Potassium magnesium sulphate (a)	18	High	
Potassium carbonate	55	high	Low palatability

<u>Notes</u>
(a) most commonly used

Potassium

20e

| K | Atomic Wt 39.10 | Atomic No 19 | Macro Mineral |

Feed Name	g/kg DM	Feed Name	g/kg DM
Alfalfa meal	2.3	Molasses - cane	39.0
Bakery waste	5.3	Oat groats	3.6
Barley grain	6.0	Oat middlings	5.5
Bean field	13.2	Oatfeed	3.9
Blood meal	1.0	Oats grain	4.8
Brewers grains	1.1	Palm kernel exp	6.6
Buckwheat grain	5.1	Peas	11.1
Buttermilk dehyd.(cattle)	9.0	Potato dried	18.6
Casein dehyd. (cattle)	0.1	Rape ext (mech)	15.6
Cassava tubers dehy	7.0	Rice bran	19.2
Citrus pulp dried	11.4	Rice grain	3.6
Copra meal	16.2	Rye grain	5.2
Cottonseed whole	12.2	Safflower ext. solv.	8.2
Cottonseed meal	13.2	Sesame ext mech	13.5
Distillers grains - wheat	1.2	Silage alfalfa	22.9
Distillers grains maize	1.8	Silage grass	18.5
Distillers grains- barley	10.3	Silage maize	7.6
Fishmeal (Sth Am)	7.7	Silage sorghum	13.7
Grass bluegrass	19.8	Silage wholecrop	8.3
Grass alfalfa	20.9	Sorghum grain	3.9
Grass bermuda	17.0	Soya ext.solv	22.5
Grass clover	22.8	Soya flour	16.9
Grass extensive	25.0	Soya hipro	23.9
Grass kikuyu	31.5	Straw barley	23
Grass timothy	19.4	Straw oat	22.5
Groundnut ext	14.3	Straw wheat	12.5
Hay alfalfa	26.7	Sugar beet pulp (dehyd)	2.0
Hay bluegrass	16.9	Sugar beet pulp (mol)	18.0
Hay clover	16.2	Sunflower Ext	9.6
Hay eragostus	9.8	Triticale grain	4.0
Hominy feed	7.0	Wheat (caustic)	5.2
Linseed meal (mech ext)	13.1	Wheat bran	15.6
Maize bran	6.8	Wheat feed	11.2
Maize germ ext(sol)	3.1	Wheat germ ext.	9.8
Maize gluten 20	9.2	Wheat grain	5.2
Maize gluten 60	1.6	Whey low lactose	31.6
Maize grain	3.5	Whey (cattle dehyd)	12.3
Malt culms	2.0	Yeast (brewers dehyd)	17.9
Milk (cattle-dehyd)	10.8	Yeast (torula dehyd)	20.4
Milk skimmed	16.0		
Millet grain	4.8		
Molasses - beet	51.0		

CONTEXT www.contextbookshop.com

Selenium

21a

| Se | Atomic Wt 78.96 | Atomic No 34 | Micro Mineral |

Introduction
- A semi metal (metalloid), similar to sulphur in its chemical properties
- Closely related to Vitamin E as a biological anti-oxidant
- Se/Vitamin E can often replace the other to an extent but not completely
- Requirements do not vary with fat intake (unlike vitamin E) but with the animal's metabolic rate and the presence in cells of active oxygen

Key Natural Sources
- Soil levels directly affect plant levels from soil status or fertilisation application
- Well-aerated and alkaline soils enhance uptake of Se by plants
- Acid soils rarely produce plants with toxic Se concentrations
- Leaves usually contain double the selenium of stems
- Soil Selenium status is suggested as Very high >1.5, High>0.9, Low <0.5, Very Low <0.3. (mg/kg air-dry soil)
- Intensive farming means many crops are deficient in selenium for modern animal production levels, (know local and bought feed levels)
- Animal Selenium -Most is associated with amino acids or proteins – rarely found in inorganic form in organs, tissues or body fluids
- Fishmeal is usually a good source

Function
- Component of enzyme glutathionine peroxidase, GSHpx, which protects against oxidative damage protecting cellular and subcellular membranes
- Acts as an antioxidant, destroys peroxides before they can attack cellular membranes. (reduces amount of vitamin E required to maintain integrity of lipid membranes)
- Redox control of cell reactions
- Protects some tissues from poisonous substances e.g. arsenics, cadmium and mercury, silver
- Involved in liver function
- Stimulates production of IgM antibody producing cells; replaces Vitamin E which enhances IgG production
- Has a role in thyroid production of thyroxine hormone and thus affects growth rates
- Essential element in the production of cysteine (an essential amino acid) from methionine
- Selenoproteins 15kd and 34kd have a preference for testicular and prostrate tissues
- Aids in retention of Vitamin E in blood plasma
- Necessary for the repair of DNA

Benefits
- Natural anti-oxidant
- Plays a role in resistance to viral infection
- Reproduction
- Growth
- Protect integrity of tissues
- Involved in and enhances immune response system
- Protects muscles from degeneration
- Required for good pancreas function (preserves integrity of pancreas)
- Readily transmissible through placenta to foetus, mammary gland and egg.
  (organic better than inorganic)
- Metabolic function closely linked with Vitamin E

Absorption
- No homeostatic control on absorption

Ruminants
- Apparent digestibility of selenium in forages and concentrates is between 35-65%
- Inorganic selenium absorbed as microbial protein. (Sodium selenite is oxidised in the rumen and utilised by rumen microbes. Microbial status may limit absorption capacity)
- Inorganic selenium most soluble in alkaline conditions
- Organic selenium metabolised by rumen microbes or absorbed in small intestine by amino acid pathway
- Apparent digestibility of selenium in sodium selenate, sodium selenite, and Se-enriched yeast range from 30-50%

Selenium

| Se | Atomic Wt 78.96 | Atomic No 34 | Micro Mineral |

Absorption

- Organic sources of selenium usually increase the level in blood and tissues more than inorganic sources

Monogastric
- Absorption mainly in the duodenum (greatest in lower small intestine, caecum, colon)
- Absorption Selenite- 30-60%, selenate –90%, selenocysteine & selenomethionine-98%

Metabolism

- Bound to a protein and transported in the blood to tissues
- Principal plasma Se (for transportation) is Selonoprotein P
- In tissues it is incorporated into tissue protein as selenocysteine and selenomethionine
- Stored in kidney, liver, spleen pancreas and muscle

- Wool and hair can have relatively high concentrations reflecting long-term dietary intake only
- Retention influenced by animal state, amount and chemical form of Se fed
- Milk and egg Se concentrations are very sensitive to dietary Se intakes

Excretion

Monogastric
- Primary routes: Urine, faeces and exhalation. (Exhalation is major route only when toxic levels are consumed.)

Ruminants
- Oral Se: most excreted by faeces (non absorbed and metabolic Se)
- Injected Se: most excreted in urine

Requirement /Allowances (mg/kg DM)

Rums	NRC	Pigs	NRC	Poultry	NRC	Others	NRC
Calf	0.3	Creep	0.3	Chick	0.15	Dog	0.11
Dairy	0.3	Weaner	0.3	Broiler	0.15	Cat	0.1
Beef	0.1	Grower	0.25	Breeder	0.06	Horse	0.1
Lamb		Finisher	0.15	Layer	0.06	Fish	0.15-0.38
Sheep	0.2	Sow/Boar	0.15	Turkey	0.2	Rabbits	

Rums	Typical	Pigs	Typical	Poultry	Typical	Others	Typical
Calf (MR)	0.3	Creep	0.3	Chick	0.3	Dog	0.11-0.3
Dairy	0.3	Weaner	0.3	Broiler	0.3	Cat	0.1-0.3
Beef	0.2	Grower	0.3	Breeder	0.3	Horse	0.3-2.0
Heifer	0.3	Finisher	0.3	Layer	0.3	Fish	
Sheep	0.3	Sow/Boar	0.3	Turkey	0.3	Rabbits	

Adequate Status

Species	Liver	Kidney	Blood	Milk	Serum
Cattle	0.25-0.5	1-1.5	19-36	0.03-0.05	0.08-0.3
Sheep	0.25-0.5	0.9-3.0	60-180		0.08-0.5
Pigs	0.4-1.2	1.5-2.5	100-200	0.12-0.2	0.14-0.3
Poultry	0.35-1.0	0.5-1.2	120-140		0.03-0.06
Horses	0.3-1.0	0.7-2.0	30-150	0.15-0.04	0.14-0.25
Rabbits	1.07-2.0	1-2			
Dogs	0.5-1.5	1-1.5			
Cats	0.26-0.54	0.77-1.14			
Units	ppm WW	ppm WW	GSH-px(μmoles /mgHb/min	ppm WW	ppm WW

Selenium

| Se | Atomic Wt 78.96 | Atomic No 34 | Micro Mineral |

Adequate Status
- Kidney and Liver are the most sensitive indicators of status
- Activity of blood GSHpx is a good indicator of status
- Typical hair levels in ppm DM are Cattle 0.5-1.3, Sheep 0.08-4.0, Pigs 0.4-2.0, Horses 1-3

Deficiency

General
- Nutritional muscular dystrophy (NMD)
- Myocardial disease
- Reproductive disorders
- Reduced disease resistance and immune response
- Poor growth rates/ feed utilisation

Dogs
- Implicated in hip dysplasia, arthritis

Cattle
- White muscle disease (newborns)
- Diarrhoea
- Muscle stiffness
- Sudden death from cardiac failure
- Retained placenta
- Cystic ovaries and metritis
- Early foetal abortions
- Weak, still born calves
- Infertility affecting oestrus, ovulation, embryo fertilisation and development
- Low sperm motility

Sheep
- White muscle disease (stiff lamb disease)
- Newborn lambs are very susceptible from birth to 3 weeks old. (Due to low ewe status and unable to pass sufficient across the placenta and via milk)
- Peridontal disease may be Se dependent

Pigs
- Mulberry heart disease
- Liver necrosis
- Low sperm motility
- Low litter size
- Low sperm motility
- Impaired spermatogenesis in boars

Poultry
- Fibrosis of the pancreas, NMD, Exudative diathesis
- Nutritional pancreatic atrophy (chicks)
- Poor brain capillary strength leading to encephalomalacia (crazy chick disease)
- Reduced egg production, hatchability
- Reduced fertility

Turkey
- Exudative diathesis, gizzard myopathy

Ducks (growing)
- Gizzard, muscular myopathies.-low growth and exudative diathesis

Horses
- Muscular dystrophy (foals)
- Azoturia (tying up)
- Anorexia, emaciation, muscle weakness, tachycardia, diarrhoea
- Early embryonic death, abortions, sudden death, Agalactica

Toxicity

General
Usually associated with:
  -Incorrect feed levels or eating selenium accumulating plants (e.g., Astragalus sp)
  -Se can accumulate in the foetus at the expense of the dam (feeds containing >5ppm Se)

Cattle
- Alkali disease and blind staggers
- Emaciation, hair loss
- Lameness, sloughing of hooves
- Anaemia
- Teeth grinding
- Blindness
- Death
- Acute toxicity(cows) 10-20 mg of Se/kg of body weight

- Chronic toxicity is nearly 20 times the normal intake

Sheep
- Blind staggers if fed astragulus species but do not show signs of alkali disease

Pigs
- Appetite loss & depressed growth
- Hair loss
- Stiffness
- Hoof separation, joint erosion
- Paralysis incoordination
- Liver cirrhosis
- Heart atrophy
- anaemia
- Impaired embryo development
- Blind staggers

Selenium

21d

| Se | Atomic Wt 78.96 | Atomic No 34 | Micro Mineral |

Toxicity

Poultry
- Reduced egg production and hatchability Increased embryo deformities
- Severe fatty metamorphosis
- Reduced weight gains
- Turkeys more resistant than chickens

Dogs
- Garlicky odour to breath, vomiting, dyspnea, titanic spasms and death (maximum dietary recommended: 2mg/kg dm)

Horses
- Blind staggers/ alkali disease

Maximum Dietary Tolerable Level (mg/kg DM)

Ruminants	2	Rabbits	2
Pigs	2	Dogs	2
Poultry	2	Horses	2

Interrelationships
- Closely related to Vitamin E, and sulphur containing amino acids in function
- Animals marginal in selenium require extra Vitamin E and vica versa
- Protects against the toxic effects of arsenic, cadmium, copper, mercury, silver, tellerium, zinc and sulphur and they also counteract the toxic effects of selenium
- High protein diets can help protect against selenium toxicity or low protein diets have higher Se requirement
- Diets with high levels of poly-unsaturated fatty acids but low in Vitamin E raise the requirement for selenium
- Riboflavin (Vitamin B_2) is involved in GSH-Px system and if deficient may aggravate Se deficiency
- Se and Vitamin E are involved in pancreatic function that produces tryptophan (soybeans contain tryptophan inhibitors)
- Genetic variations in requirements and susceptibility have been identified

Antagonists
- High levels of ferric iron in soil
- Sulphur and arsenic can limit effectiveness
- High and low dietary calcium levels reduces selenium absorption
- High linoleic acid increase incidence of WMD when Se intake is marginal

Synergy
- Vitamin E
- Ascorbic acid increases dietary Se absorption
- Adequate Vitamin E and B_6 enhance Se utilisation

Main Supplements

Source	Element %	Relative Bio-avail.	Comments
Sodium selenite Na_2SeO_3*	45.6	high	High Hygroscopicity & H_2O Solubility
Sodium selenate	40.0	high	
Calcium selenite			
Selenium dioxide			
Selenium enriched yeast	0.1-0.2	high	
Seleno methionine			
Slow release rumen bolus			

Notes
There are strict legal limits of inclusion levels and sources.
* Limited to <0.3 ppm DM in feeds

CONTEXT www.contextbookshop.com

Selenium

| Se | Atomic Wt 78.96 | Atomic No 34 | Micro Mineral |

Feed Name	mg/kg DM	Feed Name	mg/kg DM
Alfalfa meal	0.33	Molasses -cane	0.07
Bakery waste	-	Oat groats	-
Barley grain	0.30	Oat middlings	-
Bean field	0.02	Oatfeed	-
Blood meal	0.80	Oats grain	0.19
Brewers grains	0.50	Palm kernel exp	0.13
Buckwheat grain	0.20	Peas	0.22
Buttermilk dehyd.(cattle)	0.11	Potato dried	0.06
Casein dehyd. (cattle)	0.16	Rape ext (mech)	0.70
Cassava tubers dehy	0.11	Rice bran	0.44
Citrus pulp dried	0.05	Rice grain	0.11
Copra meal	-	Rye grain	0.44
Cottonseed Whole	-	Safflower ext. solv.	-
Cottonseed meal	0.15	Sesame ext mech	0.22
Distillers grains - wheat	0.16	Silage alfalfa	-
Distillers grains maize	0.18	Silage grass	0.02
Distillers grains- barley	0.40	Silage maize	0.01
Fishmeal (Sth Am)	1.75	Silage sorghum	-
Grass bluegrass	-	Silage wholecrop	0.02
Grass alfalfa	-	Sorghum grain	0.50
Grass bermuda	-	Soya ext.solv	0.28
Grass clover	-	Soya flour	-
Grass extensive	0.02	Soya hipro	0.28
Grass kikuyu	0.02	Straw barley	0.04
Grass timothy	-	Straw oat	0.03
Groundnut ext	0.14	Straw wheat	0.03
Hay alfalfa	0.33	Sugar beet pulp (dehyd)	0.03
Hay bluegrass	-	Sugar beet pulp (mol)	0.17
Hay clover	-	Sunflower ext	0.34
Hay eragostus	-	Triticale grain	-
Hominy feed	0.11	Wheat (caustic)	0.10
Linseed meal (mech ext)	0.89	Wheat bran	0.43
Maize bran	17.00	Wheat feed	0.56
Maize germ ext(sol)	0.26	Wheat germ ext.	0.40
Maize gluten 20	0.05	Wheat grain	0.10
Maize gluten 60	0.17	Whey low lactose	0.06
Maize grain	0.22	Whey(cattle dehyd)	0.06
Malt culms	0.60	Yeast (brewers dehyd)	0.98
Milk (cattle-dehyd)	0.31	Yeast (torula dehyd)	1.08
Milk skimmed	0.13		
Millet grain	0.80		
Molasses -beet	0.26		

CONTEXT   www.contextbookshop.com

Sel-Plex® from Alltech

| Se | Atomic Wt 78.96 | Atomic No 34 | Micro Mineral |

Introduction
- Sel-Plex® is a selenium yeast for use in livestock feeds.

Concentrations Available*
- 2000 ppm
- 1000 ppm
- 600 ppm

*Not all mineral concentrations are available in every country. Contact your local Alltech representative for details.

Physical Characteristics

Appearance
Sel-Plex® is a tan to brown, free flowing powder with typical yeast-like aroma.

Storage
Sel-Plex® should be stored in a cool, dry place. Open containers should be resealed. Shelf life under these conditions is 12 months.

Packaging
Sel-Plex® is available in 25 kg bags.

Sel-Plex® Inclusion Rates
Supply 100% of the selenium requirement with Sel-Plex®.

Example:
To supply 0.3 ppm selenium using Sel-Plex® (2000 ppm), the inclusion rate is 150 grams per tonne.

For inclusion recommendations, maximum permitted selenium levels for your region and registered label guidelines, contact your local Alltech representative.

Benefits (Conditions responsive to improved Selenium status)
- Selenium works with vitamin E in metabolic functions
- Promotes normal body growth, enhances fertility and encourages tissue elasticity
- A potent antioxidant that naturally reduces the retention of toxic metals in the body
- Selenium is crucial for the proper functioning of the heart muscle and there is evidence that selenium can help the body fight cancer

Some Symptoms of Selenium Deficiency
- White Muscle Disease
- Weak newborns
- Infertility
- Immune deficiencies
- Reproductive disorders
- Cancer
- Fatigue
- Fibromyalgia
- Heart disease
- Muscular weakness

www.alltech.com

Silicon

| Si | Atomic Wt 28.09 | Atomic No 14 | Minor Mineral |

Introduction
- Second most abundant element on earth next to oxygen. Comprises 25.7% of earths crust by weight
- The name 'Silicon' comes from the Latin word *silicis* which means flint
- A semi-metalloid, never found in its natural state, but in combination with oxygen as a silicate ion (SiO_4) in silica-rich rocks such as obsidian, granite, diorite, and sandstone. Feldspar and quartz are the most significant silicate minerals
- It alloys with a variety of metals, including iron, aluminum, copper, nickel, manganese and ferrochromium
- Found in large quantities within rock and therefore soil
- Most pure sand is silica dioxide
- Used in manufacture of special steels and cast iron, aluminum alloys, glass and refractory materials, ceramics, abrasives, water filtration, component of hydraulic cements, filler in cosmetics, pharmaceuticals, paper, insecticides, rubber reinforcing agent - especially for high adhesion to textiles, anti-caking agent in foods, flatting agent in paints, thermal insulator. Fused silica is used as an ablative material in rocket engines, spacecraft, silica fibres used in reinforced plastics
- In animals, found mainly in skin and its appendages e.g. the ash level of feathers shows high levels

Key Natural Sources
- Level in plant foods greater than animal feedstuffs
- Levels is plants vary according to age of maturity, species and soil type and can reach over 50 mg/kg DM
- Some organisms like sponges and some plants use silicon to create structural support
- Cereal grains, a rich source particularly more fibrous, e.g. Oats.
- Soil contamination of feeds increases intake

Function
- Involved in growth and skeletal development
- Involved in bone mineralisation. (A component of glycosaminoglycans found in the bone matrix) helps calcium in maintaining bone strength
- May be essential for connective tissue development particularly cartilage. (Appears to increase the collagen content and to be a constituent of mucopolysaccharides)
- May be a factor in the immune system

Absorption
- Solid silica appears to be dissolved in the gut and absorbed as monosilicic acid
- Organic bound silica may improve absorption
- High levels in forage for ruminants, depresses DM digestibility
- Ruminants - Poor absorption

Metabolism
- Concentration in the blood is generally just below 0.01mg per ml

Excretion
- Excreted in urine and faeces

Silicon

Si | Atomic Wt 28.09 | Atomic No 14 | **Minor Mineral**

Requirement /Allowances (mg/kg DM)

Rums	Typical	Pigs	Typical	Poultry	Typical	Others	Typical
Calf		Creep		Chick	50-250	Dog	
Dairy	5-30 g/d	Weaner		Broiler		Cat	
Beef		Grower		Breeder		Horse	
Lamb		Finisher		Layer		Fish	
Sheep	5-10g/d	Sow		Turkey		Rabbits	
Bull		Boar		Quail			

Adequate Status

Species	Liver	Urine	Serum	Milk	Feathers
Cattle	0.2-0.5		1.6	0.6-2.0	
Sheep		10-150		4.8	
Pigs					
Poultry					1.6
Units	ppm DM	mg/d DM	ppm DM	ppm DM	ppm WW

Deficiency

General
- Unlikely
- Reduce growth rate
- Hair loss
- Affect bone formation

Chicks
- Smaller long bone joints
- Reduced bone strength
- Skull abnormalities

Toxicity

Ruminants
- Si accumulates in the bladder, kidney or urethra to form calculi or uroliths, "Water belly". If develop to become large they will block the urinary passage and lead to death
- The maximal tolerable silicon concentration 0.2% of the diet

Sheep
- High pasture levels: teeth detoriation due to hard opal phylotihs of pasture plants

Monogastric
- Essentially non-toxic

Interrelationships
- Calcium, Magnesium, Aluminium and Fluorine

Main supplements
- Not supplemented to animals as so widely found

Sodium

23a

Na | Atomic Wt 22.99 | Atomic No 11 | Macro Mineral

Introduction
- The sixth most abundant element in the Earth's crust at 2.83% by weight
- Named Sodium comes from 'soda' and symbol from Latin word *'natrium'* (sodium). An important trade item in Roman times
- The second most abundant element dissolved in seawater (after chloride)
- Halite (Sodium chloride--Salt): Used in human and animal diet, food seasoning and food preservation, used to prepare sodium hydroxide, soda ash, caustic soda, hydrochloric acid, chlorine, metallic sodium, used in ceramic glazes, metallurgy, curing of hides, mineral waters, soap manufacture, home water softeners, highway de-icing, photography, herbicide, fire extinguishing, nuclear reactors, mouthwash, medicine (heat exhaustion), in scientific equipment for optical parts.
- Single crystals used for spectroscopy, ultraviolet and infrared transmission
- Sodium sulphates – used extensively in the manufacture of pulp and paper, dyes and ceramic glazes
- Sodium carbonate – used in manufacture of glass, pulp and paper, and rayon
- Sodium nitrate – an ingredient in fertilisers and explosives
- Body contains approx. 0.16% sodium
- 30 to 50% of total body sodium is found in the bone, in an insoluble form
- Largest proportion found in soft tissues and body fluids. A component of bile, pancreatic juice, sweat, tears

Key Natural Sources
- Does not occur free in nature, most important sodium salts are sodium chloride (rock salt), sodium carbonate (soda), sodium borate (borax), sodium nitrate and sodium sulphates
- Sodium salts are found in seawater, salty lakes, alkaline lakes and mineral spring water
- Sodium Chloride or Salt is the main source of sodium in the diet
- Most plant and plant products contain relatively low levels of Na compared to animal products (meat meal, fishmeal, dried skimmed milk etc)
- Plant levels affected by fertiliser application (e.g. potassium chloride depress Na uptake)
- Tropical pasture species generally accumulate less Na than temperates species

Function
- Major extracellular cation
- Major electrolyte, with K and Cl, in acid-base balance of the body
- Essential in osmotic pressure
- Essential for nutrient transfer to cells and removal of waste
- Involved in appetite(palatability)
- Involved in body water regulation
- Helps control stomach pH
- Necessary for muscle and heart contraction
- Necessary for absorption and metabolism of sugars and amino acids
- Transmission of nerve impulses
- Involved in sodium-potassium pump essential for cell transport of glucose, amino acids, and phosphate in and calcium, hydrogen, bicarbonate, potassium, and chloride ions out of cells
- Constituent of saliva salts to buffer acid from ruminal fermentation

Benefits
- Osmotic balance
- Acid –base balance
- Muscle/ cell formation
- Enables more efficient utilisation of digested protein and energy
- Absorption of several water soluble vitamins may be Na-coupled

Absorption
- Principally from upper small intestine with no apparent controls
- Absorption is very efficient (90-95%) provided glucose is available for transport purposes
- Transport across intestinal epithelium dependent on a system of 'pumps' and passive 'leaks' in cell membranes
- Ruminants, absorption also occurs in the reticulorumen, abomasum, omasum and duodenum as well as intestines
- 80% of Na + Cl entering GI tract is from internal secretions e.g. saliva, gastric fluids, bile, pancreatic juice

CONTEXT   www.contextbookshop.com

Sodium

| Na | Atomic Wt 22.99 | Atomic No 11 | Macro Mineral |

Metabolism
- Virtually no stores
- Na is carried in the blood, filtered out by the kidneys and returned to the blood at the levels required. Reasorption of sodium can occur in kidney by active transport and alterations in luminal surface permeability
- Body sodium concentrations controlled by hormones, including aldosterone and an antidiuretic hormone of the posterior pituitary, to maintain a constant ratio of Na to K in the extracellular fluid
- Intestine may be important in regulating Na balance during deficiency
- Salt appetite is an overriding force for Na and Cl intake in omnivores and herbivores
- Ruminant sodium balance may be affected by loss of saliva especially during elevated environmental temperature

Excretion
- Mainly by urine as chlorides and phosphates
- Smaller amounts lost in faeces and sweat
- Also lost from vomiting and diarrhoea
- Urinary sodium is directly correlated to Na intake. Urine Na decreases with Na deficiency but increases with Cl deficiency
- Milk contains about 630mg/l Na
- Cattle saliva contains approx 160-180 meq/l

Requirement /Allowances (g/kg DM)

Rums	NRC	Pigs	NRC	Poultry	NRC	Others	NRC
Calf	2	Creep	2.5	Chick	2	Dog	0.6
Dairy	2.0-3.4	Weaner	2	Broiler	1.5	Cat	0.5
Beef	0.6-0.8	Grower	1.5	Breeder	1.5	Horse	1-3
Heifer	0.7-0.8	Finisher	1	Layer	1.5	Fish	6
Sheep	0.9-1.8	Sow/Boar	1.5-2.0	Turkey	1.2-1.7	Rabbits	2
Rums	Typical	Pigs	Typical	Poultry	Typical	Others	Typical
Calf (a)	1	Creep	1	Chick	1.5	Dog	0.6-3
Dairy	1.3-2	Weaner	2.4	Broiler	1.8	Cat	2
Beef	0.6	Grower	2	Breeder	1.8	Horse	3
Heifer	1	Finisher	2	Layer	1.8	Fish	
Sheep	0.8-1.0	Sow/Boar	1.5-2.3	Turkey	1.7	Rabbits	

Notes
(a) Milk replacer 4-6 g/kg DM

- Lactation, pregnancy, and growth affect the requirement for sodium (salt) Factors influencing salt requirements: level of K in diet; dry vs green forage; levels in water; genetic differences for milk levels, illness or disease (diarrhoea, vomiting, renal losses); resistance to disease and parasitism, geographical location, temperature and humidity, sweating capability of species, animal class and physiological status, type of feeds.

Sodium

| Na | Atomic Wt 22.99 | Atomic No 11 | Macro Mineral |

Adequate Status

Species	Liver	Serum	Urine	Brain
Cattle	900-1800	135-150	8-40	
Sheep	2mg/gdw	140-152	1.2-1.9g/d	
Pigs		135-150	60-250ppm	1850-2030
Poultry		131-142		1600-1710
Horses		130-143		
Rabbits		110-156		
Dogs	1200-1700	141-148		1200-1600
Cats		141-155		
Units	ppm WW	mEq/l DM	mEq/l DM	ppb WW

Deficiency

General
- Poor feed conversion
- Reduced reproduction efficiency through poor fertility in males
- Loss of appetite and weight loss
- Lack lustre eyes/ poor coat
- Reduced growth/ ill thrift

Cattle
- Pica, depraved appetite (salt craving)
- Reduced milk production in extreme deficiency
- In co-ordination
- Shivering
- Dehydration
- Arrhythmia of the heart

Sheep
- Depraved appetite, continual bleating

Pigs
- Reduced water consumption
- Reduced FC and weight gain

Poultry (Chicks)
- Reduced egg production and size
- Increased cannibalism and moulting
- Weak or abnormal bones
- Corneal keratinisation
- Reduced hatchability
- Possible reduced immune response
- Serum Ca and Mg increase with NaCl deficiency

Dogs
- Fatigue, exhaustion, inability to maintain water balance, decreased water intake, retarded growth, dry skin, loss of hair

Horses
- Lick mangers, fences, dirt etc
- Severe (from sweating) fatigue and exhaustion

Toxicity

General
- Very much affected by the availability of water
- Or animals have been used to a very low salt diet and it has changed
- Or animal exposed to high salt/sodium feed
- Creates hypertension

Ruminants
- Rarely seen until > 1% of DM for cattle
- Gastrointestinal irritation
- Reduced efficiency of organic matter utilisation, weight loss
- Central nervous system impairment
- Diarrhoea, vomiting, excessive thirst, salivation, staggering.
- Blindness, Oedema

Poultry
- Ascites
- Excessive water intake/wet faeces
- Increased mortality
- Reduced egg production, feed efficiency and egg weight
- Reduced growth in males
- Sudden death
- May reduce egg fertility

Dogs
- Relatively resistant due to efficient excretion of Na ions via kidney
- Ataxia, incoordination, blindness, epileptiform seizures and convulsions
- Severe gastritis

Sodium

| Na | Atomic Wt 22.99 | Atomic No 11 | Macro Mineral |

Maximum Dietary Tolerable Level of NaCl (g/kg DM)

Lactating Cattle	40	Rabbits	30
Non Lactating Cattle	90	Horses	30
Sheep	90	Poultry	20
Pigs	80	Chicks	>9
		Layers	>12

Maximum Dietary Tolerable Level of Na (g/kg DM)

Lactating Cattle	15.7	Rabbits	11.8
Non Lactating Cattle	35.4	Horses	11.8
Sheep	35.4	Poultry	7.86
Pigs	31.4	Chicks	3.53
		Layers	4.7

Interrelationships
- Closely related in the animal's metabolism to potassium and chlorine
- Haemoglobin, urea N, ornithine, lysine and total basic amino acids decrease linearly as dietary Na increases. (pigs)
- Na may remove excess P from blood thus increasing eggshell quality. (poultry)

Antagonists
- Potassium (excess aggravates Na deficiency)
- Heat stress

Synergy
- Monensin and laslaocid in high concentrate diets reduces Na retention
- Dry forages require less voluntary intake than green forage

Main Supplements
- Na and Cl most often limiting if not supplemented
- Poultry: Salt supplement minimised to reduce moisture level in excreta
- Buffers (i.e) sodium bicarbonate are added to rations to reduce digestive upset

Main Supplements

Source	Element %	Relative Bio-avail.	Comments
Sodium bicarbonate	27	high	
Sodium chloride (salt)	40	high	
Sodium phosphate dibasic		high	
Sodium phosphate monobasic		high	
Sodium sulphate		high	
Sodium acetate		high	
Monosodium glutamate			
Sodium aluminosilicate		high	

Sodium

Na | Atomic Wt 22.99 | Atomic No 11 | Macro Mineral

Feed Name	g/kg DM	Feed Name	g/kg DM
Alfalfa meal	0.7	Molasses - cane	2.4
Bakery waste	12.4	Oat groats	0.6
Barley grain	0.35	Oat middlings	1.0
Bean field	0.2	Oatfeed	0.1
Blood meal	3.5	Oats grain	0.7
Brewers grains	0.6	Palm kernel exp	0.7
Buckwheat grain	0.6	Peas	0.3
Buttermilk dehyd. (cattle)	9.0	Potato dried	0.2
Casein dehyd. (cattle)	0.1	Rape ext (mech)	0.4
Cassava tubers dehy	0.3	Rice bran	0.4
Citrus pulp dried	0.5	Rice grain	0.6
Copra meal	0.4	Rye grain	0.3
Cottonseed whole	3.2	Safflower ext. solv.	0.5
Cottonseed meal	0.2	Sesame ext mech	0.4
Distillers grains - wheat	3.1	Silage alfalfa	1.5
Distillers grains maize	1.2	Silage grass	2.4
Distillers grains- barley	1.2	Silage maize	0.4
Fishmeal (Sth Am)	9.8	Silage sorghum	0.2
Grass bluegrass	1.6	Silage wholecrop	0.3
Grass alfalfa	1.9	Sorghum grain	0.3
Grass bermuda	0.1	Soya ext.solv	0.2
Grass clover	2.0	Soya flour	2.8
Grass extensive	1.7	Soya hipro	0.2
Grass kikuyu	4.3	Straw barley	2.2
Grass timothy	1.4	Straw oat	3.8
Groundnut ext	0.4	Straw wheat	0.5
Hay alfalfa	1.1	Sugar beet pulp (dehyd)	2.1
Hay bluegrass	1.3	Sugar beet pulp (mol)	4.5
Hay clover	1.9	Sunflower ext	0.5
Hay eragostus	0.2	Triticale grain	0.3
Hominy feed	0.4	Wheat (caustic)	28.0
Linseed meal (mech ext)	1.2	Wheat bran	0.4
Maize bran	0.2	Wheat feed	0.5
Maize germ ext(sol)	0.5	Wheat germ ext.	0.3
Maize gluten 20	1.6	Wheat grain	0.3
Maize gluten 60	0.6	Whey low lactose	15.4
Maize grain	0.3	Whey (cattle dehyd)	7.0
Malt culms	0.6	Yeast (brewers dehyd)	0.8
Milk (cattle-dehyd)	3.8	Yeast (torula dehyd)	0.4
Milk skimmed	5.3		
Millet grain	0.5		
Molasses -beet	13.0		

CONTEXT www.contextbookshop.com

Sulphur

24a

| S | Atomic Wt 32.06 | Atomic No 16 | Macro Mineral |

Introduction
- One of most abundant elements in nature
- A bright, lemon yellow, soft, non-metallic element.
- Has a very low thermal conductivity; melts at 108°C, and burns easily with a blue flame
- S gas combined with oxygen produces sulphur dioxide, which smells like rotten eggs
- Found in large quantities in rock from volcanic regions
- Mined sulphur is mostly from salt domes or bedded deposits. The vast majority is produced as a by-product of oil refining and natural gas processing
- Gypsum, calcium sulphate is an important source
- Name 'Sulphur' comes from the Latin word, *'sulphurium'* which means "burning stone"
- Used in the manufacture of sulphuric acid, fertilisers, chemicals, explosives, dyestuffs, petroleum refining, vulcanisation of rubber, fungicides, steel processing, bleaching agents for paper pulp, sugars, vegetable oils, food and drink preservation
- Used in various medicines, particularly sulpha drugs or sulphonamides that fight bacteria and other organisms
- These drugs prevent bacteria from multiplying
- Found in every cell in the body and makes up approx 0.15-0.25% of the body or 10% of the mineral content
- A key component of amino acids (e.g. Cystine, Cysteine and Methionine). All living animals require methionine
- Part of two vitamins- Biotin and Thiamine

Key Natural Sources
- Found widely in the air but levels in developing markets are reducing as pollution gets cleaned up
- Feed values vary depending and proportional to the protein level
- Soil levels vary widely and availability may be reduced in some clays as the sulphur is bound up
- Soil of tropics generally lower S compared to temperate regions
- Burning of dry grasses reduces S levels by volatisation
- Various forages have different S requirements for optimum yields
- High S fertilisation increases levels in forage
- S bioavailability varies with type of forage
- In raw materials the total sulphur including amino acids must be considered
- Maize gluten meal, molasses and brassicas, fish, feather, meat and blood meals are high sources of S
- Non-ruminants can obtain S only from organic sources
- Ruminants can obtain S from inorganic or organic sources
- Organically bound sulphur is more available than sulphates to most livestock
- A major source of S intake can be from sulphates in water

Function
- Component of sulphur containing amino acids e.g. cystine, cysteine, methionine and taurine
- Cystine and methionine required to make milk protein
- Detoxifier e.g. it rids the body of phenols and cresols
- Involved in carbohydrate metabolism through being a component of thiamine and insulin
- Involved in connective tissue as it is a component of complex carbohydrates, (Glucosamine and mucopolysaccharides (including chondroitin sulphate) are found in the cartilage and bones)
- Involved in energy metabolism through being a component of Coenzyme A
- Involved in fat metabolism as a component of Biotin
- Regulates energy through being a component of insulin and glutathionine
- Also part of haemoglobin, cytochromes, coenzyme M, lipoic acid, heparin, metallothionein, hormone-oxytocin
- Important for the manufacture of microbial protein
- Stimulates synthesis of riboflavin and vitamin B_{12} in rumen

CONTEXT

Sulphur

24b

| S | Atomic Wt 32.06 | Atomic No 16 | Macro Mineral |

Benefits
- Ruminants, horses and possibly rabbits can use elemental sulphur to synthesise sulphur amino acids through their microorganisms.
- This is optimal when fermentable energy, degradable S, N and phosphorus are supplied at levels to match the synthetic capacity of the microbial biomass

Absorption
- Occurs in rumen (through ruminal wall) and small intestine (inorganic S, by active transport)
- Sulphide absorption across ruminal wall is pH dependent
- Absorption is very efficient

Metabolism
- Sulphur containing amino acids are split off from other proteins during digestion and taken in the portal circulation
- All S compounds can be synthesised 'in vivo' from methionine with exception of thiamine and biotin
- Monogastric animals rely on S-amino acid sources for organic S requirements

Ruminants
- S and natural S compounds must be oxidised to sulphate or reduced to sulphide before use for biosynthetic reactions
- Largely a microbial process
- Ruminant microorganisms can synthesise S-amino acids, thiamine and biotin from inorganic S sources
- Sulphur accumulates in wool, hair, feather, horn, bone cartilage and muscles, but it is found in every cell in the body

Excretion
- Present in saliva as both inorganic and organic forms
- Higher levels found in cattle saliva than sheep. Can be recycled to rumen
- Excess is lost in the urine (principally) and faeces
- Urinal S forms: inorganic sulphate (from oxidation of organic S); ethereal S (from complex detoxication products); neutral S (occurs as cystine, taurine, thiosulphates etc)
- Faecal S mainly from inorganic sources

Requirements
- All animals require supplementation either as constituent of sulphur amino acids or as sulphates
- Monogastrics require organic S sources in diet
- Ruminants have S requirement for ruminal microbes and the animal
- Anaerobic fungi may be sensitive to S level in rumen
- Ruminants: where high levels of NPN are fed, additional sulphur is essential for its utilisation. (1part S to 15 parts NPN)

Sulphur

24c

| S | Atomic Wt 32.06 | Atomic No 16 | Macro Mineral |

Requirement /Allowances (g/kg DM)

Rums	NRC	Pigs	NRC	Poultry	NRC	Others	NRC
Calf (a)		Creep	-	Chick	-	Dog	-
Dairy	2	Weaner	-	Broiler	-	Cat	-
Beef	1.5	Grower	-	Breeder	-	Horse	1.5
Lamb	1.6	Finisher	-	Layer	-	Fish	-
Sheep	1.4-2.6	Sow/Boar	-	Turkey	-	Rabbits	-
Rums	Typical	Pigs	Typical	Poultry	Typical	Others	Typical
Calf (a)	2	Creep		Chick		Dog	5
Dairy	1.7-2.4	Weaner		Broiler		Cat	11
Beef	1.5	Grower		Breeder		Horse	1.5
Heifer	1.6	Finisher		Layer		Fish	
Sheep	1-2	Sow/Boar		Turkey		Rabbits	5.5

Notes
(a) 2.9 for milk replacer

Adequate Status

- Best indicator is dietary S content and animal performance
- Ruminants: Ruminal fluid can determine if adequate sulphide for microbial protein synthesis
  No reported tissue levels

Deficiency

General
- Slower growth as insufficient sulphur amino acids for the requirements of protein synthesis

Ruminants
- Rumen microbes have a decreased ability to function normally. Reduced digestibility (e.g. cellulose) of feeds and N retention
- Loss of appetite, reduced rumen fermentation, reduced weight gain, weepy, cloudy eyes, apathy and weakness, emaciation, death
- Hair shedding
- Methionine synthesis affected
- Affects utilisation of NPN.
- Microbial population does not utilise lactate and so it accumulates in rumen, blood and urine

Pigs
- Reduced performance and potential hoof problems
- Reproductive performance may be affected

Poultry
- Rare if S-amino acid levels are adequate
- Methionine is the first limiting and its deficiency will reduce growth, egg output and size and feathering

Sheep
- Poor wool growth and shedding

Sulphur

24d

| S | Atomic Wt 32.06 | Atomic No 16 | Macro Mineral |

Toxicity

General
- Dependant on form and route of administration
- Tolerance partly reflects different rate of ingestion, absorption and capture (rumen microbes)
- Excess S-containing amino acids: growth can be reduced. Tissue damage may also occur leading to hypoglycaemia
- Excess inorganic S can reduce growth and cause bone malformations also potentially leading to gastroenteritis
- Excess inorganic sulphate toxic due to conversion to hydrogen sulphide by GI flora
- Margin between requirement and toxic levels is extremely small.
- Dehydration, tachycardia, diarrhoea, metabolic acidosis, hypokalemia, polioencephalomalcia

Ruminants
- Production of hydrogen sulphide in the rumen, which is toxic to rumen microorganisms
- Reduced feed intake, ruminal motility. Thrashing, kicking at stomach, staggering, moaning, death

Pigs
- Hydrogen sulphide toxicity less due to limited ability of microflora.
- Sodium sulphate toxicity similar signs to salt toxicity if deprived of water

Poultry
- High levels in water can interfere with some vaccination programs.
- Mg sulphate more toxic than Na sulphate

Horses
- Dull, lethargic, colic, yellow froth from nostrils, increasing respiratory difficulty. Mucous membranes pale and jaundiced leading to red flush of mucous membranes. Liver dysfunction

Maximum Dietary Tolerable Level (ppm DM)

| Cattle | 3000-4000 | Sheep | 4000 |

Interrelationships
- Closely associated with N, Cu, Mo and Se

Antagonists
- With Mo interferes with copper metabolism (forms insoluble Cu-Mo-S (thiomolybdate) complexes)
- Sulphate influences Mo excretion in urine and level in blood
- Formation of insoluble copper sulphides in rumen, reduces Cu absorption
- Se interferes with enzymes of S metabolism (compete for reactive sites)
- S influences metabolism of selenium by rumen microbes
- High sulphates can interfere with Se by potentially inhibiting its uptake, reducing its availability or replacing it in key materials
- High S fertiliser can reduce Se content of plants
- High Zn increases faecal S
- P retention is reduced by sulphate in presence of high dietary Cu in lambs

Synergy
- Dietary N:S ratio is important for optimum utilisation of S by rumen microorganisms
- N:S ratios : beef, lamb, milk 15:1; wool 5:1; plants 13:1
- Approximate desirable N:S ratio: sheep 10-13.5:1; cattle 13.5-15:1
- S related to the amino acids: methionine, cystine, cysteine also to biotin, thiamine coenzyme A, glutathione and complex carbohydrates
- Methionine: choline: sulphate inter-relationship important in poultry

CONTEXT   www.contextbookshop.com

Sulphur

24e

| S | Atomic Wt 32.06 | Atomic No 16 | Macro Mineral |

Main supplements

- Usually supplied via sulphur containing amino acids, methionine and cysteine
- Trace mineral supplements such as copper sulphate, zinc sulphate, cobalt sulphate, manganese sulphate provide source of S
- Rumen-protected S-amino acids provide a dietary source to meet high production requirements and where high methionine is required, e.g wool production

Main Supplements

Source	Element %	Relative Bio-avail.	Comments
Ammonium sulphate	24	high	
Calcium sulphate (gypsum)	12-20.1	high	
Sodium sulphate	10	high	Not permitted *
Potassium sulphate	18	high	Not permitted *
K,Mg sulphate	22	high	
DL-methionine	20	high	
Sulphur (flowers of)	96	low	

<u>Notes</u>

* Not permitted in livestock feeds

Sulphur

| S | Atomic Wt 32.06 | Atomic No 16 | Macro Mineral |

Feed Name	g/kg DM	Feed Name	g/kg DM
Alfalfa meal	1.9	Molasses - cane	5.0
Bakery waste	0.2	Oat groats	2.2
Barley grain	1.6	Oat middlings	2.4
Bean field	3.3	Oatfeed	1.0
Blood meal	3.7	Oats grain	2.2
Brewers grains	3.0	Palm kernel exp	3.0
Buckwheat grain	1.6	Peas	2.2
Buttermilk dehyd. (cattle)	0.9	Potato dried	0.9
Casein dehyd. (cattle)	6.5	Rape ext (mech)	7.8
Cassava tubers dehy	5.7	Rice bran	2.0
Citrus pulp dried	3.0	Rice grain	0.5
Copra meal	3.6	Rye grain	1.7
Cottonseed whole	2.6	Safflower ext. solv.	1.4
Cottonseed meal	5.0	Sesame ext mech	3.5
Distillers grains - wheat	4.04	Silage alfalfa	3.6
Distillers grains maize	4.3	Silage grass	1.6
Distillers grains- barley	3.8	Silage maize	0.8
Fishmeal (Sth Am)	5.0	Silage sorghum	1.1
Grass bluegrass	2.9	Silage wholecrop	1.1
Grass alfalfa	3.7	Sorghum grain	1.5
Grass bermuda	2.1	Soya ext.solv	5.0
Grass clover	1.7	Soya flour	0.7
Grass extensive	2.0	Soya hipro	5.0
Grass kikuyu	3.0	Straw barley	1.7
Grass timothy	1.3	Straw oat	2.3
Groundnut ext	10	Straw wheat	1.4
Hay alfalfa	4.0	Sugar beet pulp (dehyd)	2.2
Hay bluegrass	1.6	Sugar beet pulp (mol)	6.0
Hay clover	1.7	Sunflower ext	4.0
Hay eragostus	1.3	Triticale grain	1.7
Hominy feed	1.1	Wheat (caustic)	1.1
Linseed meal (mech ext)	4.0	Wheat bran	2.4
Maize bran	0.8	Wheat feed	2.0
Maize germ ext(sol)	3.0	Wheat germ ext.	2.8
Maize gluten 20	2.7	Wheat grain	1.9
Maize gluten 60	7.0	Whey low lactose	11.5
Maize grain	1.1	Whey (cattle dehyd)	11.2
Malt culms	6.7	Yeast (brewers dehyd)	4.5
Milk (cattle-dehyd)	3.2	Yeast (torula dehyd)	5.9
Milk skimmed	3.4		
Millet grain	1.5		
Molasses -beet	5.7		

www.contextbookshop.com

Tin

Sn | **Atomic Wt 118.29** | **Atomic No 50** | **Minor Mineral**

Introduction
- A silvery-white metallic element.
- The name 'Tin' is an ancient Anglo-Saxon word. Sn, comes from the Latin word for tin, *'stannum'*
- Tin is malleable (easily shaped by hammering) and has a relatively low melting point, easy to melt and cast in a clay mold
- It is stable in air and water, does not oxidise or react easily
- Tin is an essential trace element, but its specific biological actions in vivo remain largely unknown
- Commonly found in: Tin-plated containers, solder, alloys, as a stabilising agent for plastics, agricultural and general fungicides, bactericides, anthelmintics, Miticides, Rodent repellents, Antifoulant marine paints
- An alloy of tin and niobium has proven to be a "superconducting" compound at very low temperatures

Key Natural Sources
- Primarily ore cassiterite and tinstone (SnO_2)
- Found in trace amounts in tissues and feeds

Function
- May contribute to the tertiary structure of proteins or other important macromolecules such as nucleic acids
- May be an oxidation-reduction catalyst and function at active site of metaloenymes

Absorption
- Inorganic Sn, very poor absorption
- Organic Sn biologically more available than inorganic

Requirement /Allowances (mg/kg DM)
3.0 mg/kg dry matter meets all dietary requirements

Deficiency

Rats
- Poor growth, decreased efficiency and food utilisation, alopecia, depressed response to sound and changes in mineral concentrations in various organs

General
- Not often seen under practical farm conditions

Toxicity

Inorganic
Toxicosis potential low but leads to:
- Anorexia
- Growth depression
- Impairs hematopoesies
- Alters Ca Metabolism
- Pancreatic, hepatic and renal lesions

Organic
- Inflammation of biliary tracts oedema of C.N.S

Chicks
- 200ppm is not toxic

Laying Hens
- 1000ppm Sn Oxide –no effect on egg production, feed intake, bodyweight

Ducks
- 1000ppm Sn may induce Se/ vit E deficiency

Mice
- 300ppm in drinking water decrease the compressive strength of the femoral bone

Main supplements
- Supplementation not usually required

Vanadium

26a

| V | Atomic Wt 50.94 | Atomic No 23 | Minor Mineral |

Introduction
- Less well known but is an essential trace element for livestock
- Widely distributed in nature
- Name 'Vanadium' comes from 'Vanadis', the goddess of beauty in Scandinavian mythology
- Plants: found at higher levels in seeds than leaves
- Animals: found in the bones, kidneys, livers and lungs
- Is an essential element to ascidians, also known as sea squirts. They concentrate vanadium in their bodies to a level one million times higher than the concentration of vanadium in seawater

Key Natural Sources
- V forms up to 110-150ppm of earths crust (prevalence equal to Cu, Pb, Zn.)
- Used to make metal alloys, (mostly steel alloys) for tools and construction purposes. E.g. alloyed with iron to make carbon steel, high-strength low-alloy steel, full alloy steel, and tool steel. Also used in dyes and mordants and as target material for X-rays
- Occurs in some phosphate supplements and some fossil fuels and jet fuels
- Poorly absorbed from soils by most plants
- Urban area air contains 2-3 times that in rural areas
- Soil ingestion is a major source for grazing animals
- Fats &oils, fresh fruits, vegetables: <1-5ppb
- Whole grains, seafood, meat & dairy 5-30ppb
- Red clover/Ryegrass: <30-160ppb

Function
- Acts as a catalyst for specific body functions
- A component of metallo-enzymes
- Incorporated into plasma transferrin (like Fe)
- Suggested role in regulation of Na, K-ATPase and other enzymes
- Possible role in hormone metabolism, glucose metabolism, lipid metabolism
- Involved in the mineralisation of teeth and bones

Benefits
- Essential for growth and reproduction
- At supplemental levels may inhibit cholesterol production in blood vessels in young animals and in older animals
- A higher supply may lower blood lipid levels
- May have therapeutic potential in diabetes

Absorption
- Vanadium is poorly absorbed by plants and animals

Metabolism
- Excess accumulates in the liver but is excreted rapidly upon reduction of intake
- Approx 5% is incorporated into milk lactoferrin

Excretion
- Rapid clearance via urine

CONTEXT   www.contextbookshop.com

Vanadium

26b

| V | Atomic Wt 50.94 | Atomic No 23 | Minor Mineral |

Requirement /Allowances (mg/kg DM)

Rums	Typical	Pigs	Typical	Poultry	Typical	Others	Typical
Calf		Creep		Chick		Dog	
Dairy		Weaner		Broiler	0.1-3.0	Cat	
Beef		Grower		Breeder		Horse	
Lamb		Finisher		Layer		Fish	
Sheep	0.1-10	Sow/Boar		Turkey		Rabbits	

Adequate Status

Species	Liver	Kidney	Blood	Milk	Serum
Cattle	0.006-0.007			0.1-0.2	1.2-1.4
Sheep	0.1-.022DM	0.2-0.47DM			
Poultry	0.018-0.038	0.1-0.18			
Ducks	0.13-0.08	0.0007-0.002			
Dogs	0.03-0.06	0.03-0.05	0.02-0.03		
Units	ppm WW	ppm WW	ppm WW	ppb DM	mg/l DM

Deficiency

General
- Not often seen under practical farm conditions as present in the air
- Young animals: reduced growth
- Reduced conception and increased abortions

Ruminants
- Reduce milk yield and lower solids
- Off spring maybe prone to sudden cramps leading to death

Goats
- Increased rate spontaneous abortion
- Reduced milk production
- Increased creatine and B-lipoprotein
- Decreased serum glucose
- Skeletal deformities

Poultry
- Reduced feather growth
- Reduce body growth rate
- Impaired reproduction
- Reduced survival of young
- Retarded bone development in chicks
- Altered red blood cell levels and Fe metabolism
- Altered blood lipid levels

Toxicity

General
- Relatively toxic – threshold 10-20ppm of diet

Ruminants
- Due to enzyme inhibition ie Na-K ATPases
- Reduced rumen DM digestibility and cellulose digestion
- Calves/lambs: diarrhoea, refusal to eat, dehydration, emaciation, dry hair coat, depressed growth, ataxia. Mortality

Pigs
- Severe growth depression, enteritis, cystitis and mortality

Poultry
- Poor plumage, reduced growth rates, increased protein in the carcasses and reduced fat. Egg quality deteriorates and nutrient absorption may also be affected
- Recommended that dietary level is below 3-5 mg/kg of feed

Vanadium

| V | Atomic Wt 50.94 | Atomic No 23 | Minor Mineral |

Maximum Dietary Tolerable Level (ppm DM)

Ruminants	50	Rabbits	10
Pigs	10	Horses	10
Poultry	10		

Synergy
- Chloride products
- Higher copper levels can help reduce the effects of excess Vanadium
- Ascorbic acid protects high dietary level in ducks
- EDTA can act as an antidote in toxicosis cases

Antagonists
- Iodide products make toxicity worse
- High Vanadium may affect some B vitamin usage

Main Supplements
- Supplementation is not advisable as sufficient from the air and feeds

Main Supplements

Source	Element %	Relative Bio-avail.	Comments
Sodium orthovandate	27%		
Vanadyl sulphate	25%		
Vanadyl acetate			
Calcium orthovandate			Natural contaminant of P

Vanadium

26d

| V | Atomic Wt 50.94 | Atomic No 23 | Minor Mineral |

Feed Name	μg/kg DM	Feed Name	μg/kg DM
Alfalfa meal	-	Molasses - cane	-
Bakery waste	-	Oat groats	-
Barley grain	200.0	Oat middlings	-
Bean field	225.0	Oatfeed	-
Blood meal	-	Oats grain	250.0
Brewers grains	-	Palm kernel exp	-
Buckwheat grain	-	Peas	-
Buttermilk dehyd.(cattle)	-	Potato dried	-
Casein dehyd. (cattle)	-	Rape ext (mech)	-
Cassava tubers dehy	-	Rice bran	-
Citrus pulp dried	-	Rice grain	25.0
Copra meal	-	Rye grain	-
Cottonseed whole	-	Safflower ext. solv.	-
Cottonseed meal	-	Sesame ext mech	-
Distillers grains - wheat	-	Silage alfalfa	-
Distillers grains maize	-	Silage grass	150.0
Distillers grains- barley	-	Silage maize	-
Fishmeal (Sth Am)	-	Silage sorghum	-
Grass bluegrass	-	Silage wholecrop	-
Grass alfalfa	-	Sorghum grain	-
Grass bermuda	-	Soya ext. solv	-
Grass clover	-	Soya flour	-
Grass extensive	120.0	Soya hipro	-
Grass kikuyu	-	Straw barley	-
Grass timothy	-	Straw oat	-
Groundnut ext	-	Straw wheat	-
Hay alfalfa	-	Sugar beet pulp (dehyd)	-
Hay bluegrass	-	Sugar beet pulp (mol)	-
Hay clover	-	Sunflower ext	-
Hay eragostus	-	Triticale grain	-
Hominy feed	-	Wheat (caustic)	200.0
Linseed meal (mech ext)	-	Wheat bran	-
Maize bran	-	Wheat feed	-
Maize germ ext(sol)	-	Wheat germ ext.	-
Maize gluten 20	75.0	Wheat grain	200.0
Maize gluten 60	-	Whey low lactose	-
Maize grain	-	Whey (cattle dehyd)	-
Malt culms	-	Yeast (brewers dehyd)	-
Milk (cattle-dehyd)	-	Yeast (torula dehyd)	-
Milk skimmed	1200.0		
Millet grain	-		
Molasses - beet	-		

CONTEXT   www.contextbookshop.com

Zinc

27a

| Zn | Atomic Wt 65.37 | Atomic No 30 | Micro Mineral |

Introduction
- A blue-gray, metallic element.
- At room temperature, zinc is brittle, but it becomes malleable at $100°C$
- Name 'Zinc' derivation unknown but comes from German word *'zinker'* that is used for element zink
- Zinc is mined in about 40 countries with China the leading producer
- Has been used for industrial, utilitarian and ornamental purposes for 2000 years
- It is relatively resistant to corrosion in air or water, so used as a protective layer on iron and steel products from rusting. (galvanising)
- Used as an alloy with copper to make brass
- Used in manufacture of paint, chemicals, in the rubber industry, in TV screens, electroplating, metal spraying, automotive parts, fluorescent lights, electrical fuses, anodes, dry cell batteries, pennies, printing inks, fungicides, nutrition, cable wrappings, engraves' plates, organ pipes, catalysts, in photocopying, in medicine
- Essential to healthy life of humans and animals: Present in the body at approximately 25mg/kg body weight
- Found in an organic protein complex within the blood. The highest concentrations are found in epidermal tissues (20% of total body zinc)

Key Natural Sources
- Found in ores principally as the sulphide
- Large regions of zinc-deficient soils exist in many countries
- Soil pH affects uptake (higher soil pH, decreases Zn availability)
- Normal soils contain 30-300ppm Zn.
- Industrial pollution, galvanised pipes etc. increase Zn in both water and plant sources
- Topical zinc application to plants increases content
- Soils near roads and highways can be contaminated with Zn from tires and motor emissions
- Plant species vary widely in Zn concentration. Generally, content declines as plant matures
- Whole grains, relatively rich in Zn but most contained in bran and germ
- Animal products contain highest levels

Function
- Primary role in enzymes as part of the molecule and as an activator
- Largely involved in the synthesis and metabolism of proteins, carbohydrates and nucleic acids.
- Part of cytosol superoxidase dismutase (as is copper)
- Needed for bone calcification
- Involved in wound healing
- Essential role in the immune system; is involved in cell membrane structure and function. (SOD in cytosol)
- Involved in the production, storage and secretion of individual hormones. (e.g.. component of the hormone thymosin produced by thymic cells that regulates cell mediated immunity)
- Key constituent of insulin, testosterone and adrenal corticosteroids
- Involved in vitamin A metabolism. (critical for mobilisation of Vitamin A from liver)
- Some micro-organisms in the rumen require Zn for growth

Benefits
- Essential for skin, bones, hair and feathers
- Affects growth, development, bone and blood formation
- Essential for the development and functioning of reproductive organs
- Improved male fertility
- Needed for transfer of carbon dioxide in red blood cells
- Is involved in prostaglandin synthesis
- Has an antioxidant effect in protecting membranes

CONTEXT   www.contextbookshop.com

Zinc

| Zn | Atomic Wt 65.37 | Atomic No 30 | Micro Mineral |

Absorption
- Dietary absorption ranges from 15-60% (highest in young animals)
- Generally absorbed throughout small intestine of monogastrics, greatest in duodenum
- Absorbed as free ions or as a complex with amino acids
- Transport across the brush border appears to be a carrier-mediated process
- Transfer within the mucosal cell is regulated by metallothionein,(liver protein)
- Absorbed zinc is regulated by zinc content of the diet and that entering the intestinal mucosal cell
- Phytin, enhanced by high Ca, decreases absorption of Zn

Metabolism
- Required daily as utilisation and excretion rapidly deplete body stores
- Metallothionein plays a central role in Zn homeostasis
- Absorbed zinc combines with plasma protein for transport to the tissues
- Zinc in plasma is bound to albumin (2/3rds) or high molecular weight proteins
- Most is deposited in the bones which is not freely available
- Zinc is stored in the mucosal cells of the gut, liver, kidney, pancreas & spleen but capacity is limited
- The available zinc pool is small hence deficiency appears quickly

Excretion
- Most lost via faeces via pancreatic, bile and other intestinal secretions Feeding pharmacological levels causes environmental concern of manure on soil
- A little loss by the urine

Requirement /Allowances (mg/kg DM)

Rums	NRC	Pigs	NRC	Poultry	NRC	Others	NRC
Calf	30 (a)	Creep	100	Chick	40	Dog	50
Dairy	45-75	Weaner	80	Broiler	40	Cat	50
Beef	30	Grower	60	Breeder	35	Horse	40
Lamb		Finisher	50	Layer	35	Fish	15-30
Sheep	20-33	Sow/Boar	50	Turkey	40-75	Rabbits	

Rums	Typical	Pigs	Typical	Poultry	Typical	Others	Typical
Calf	60 (b)	Creep	150	Chick	60	Dog	80-120
Dairy	40	Weaner	120*	Broiler	80	Cat	75
Beef	50	Grower	100*	Breeder	80	Horse	50-100
Lamb		Finisher	50-120	Layer	50	Fish	50
Sheep	50	Sow/Boar	80-120	Turkey	80	Rabbits	

Notes
(a) Milk replacer fed - 40mg/kg DM
(b) Milk replacer fed - 65mg/kg DM
* zinc addition of 2000-3000 ppm Zn as ZnO shown to improve growth performance.
EU regulation: maximum inclusion in feedstuffs for producing animals: Poultry, Pig, Beef; 150mg/kg; Dairy: 200mg/kg.

Zinc

| Zn | Atomic Wt 65.37 | Atomic No 30 | Micro Mineral |

Adequate Status

Species	Liver	Kidney	Serum	Pancreas	Hair
Cattle	25-100	18-25	0.8-1.4	25-50	100-150
Sheep	30-75	20-40	0.8-1.2	18-30	
Pigs	40-90	15-30	0.7-1.5	35-40	150-230
Poultry	25-40	22-32	1.85-3.4	50-125	
Horses	40-125	20-50	0.6-1.7		
Rabbits	30-80	10-30	1.7-2.1		
Dogs	30-70	16-30	0.7-2.0		150-250
Units	ppm ww	ppm ww	ppm ww	ppb ww	ppm DM

- Relatively insensitive, particularly in early stages of deficiency
- Determination of dietary content but monogastrics limited by avail ability effects of Ca and phytate
- Blood Zn levels inversely related to milk production. Zinc content of milk 2.3-7.5mg/l
- Serum levels affected by disease, stress, hormones, pregnancy, starvation, protein plasma status
- Hair may be useful for long term monitoring
- Foetal liver levels increase with gestation
- Pancreas levels probably most reliable

Deficiency

General
- Reduced growth rates and poor skin condition
- Loss of appetite
- Bone problems
- Poor hair formation and slipping of wool
- Delays healing of wounds
- Poor testicular development
- Impairment of glucose tolerance
- Affects immune system
- Parakeratosis (thickening of the epithelial cells of the skin)
- Emaciations
- Impairs sexual function, reduced conception rate, severely impaired spermatozoan maturation
- Deranged electrolyte balance
- Offspring can suffer from malformations and behavioural problems

Ruminants
- Skin, hair and wool problems – sheep can show breaks in wool
- Hoof and horn weakened
- Parakeratosis of skin on legs, neck and head
- Calves bowing of hind legs and stiff joints
- Reduced feed intake
- Reduced growth rate
- Testes show reduced growth

Pigs
- Parakeratosis
- Reduced conception rate
- Increased incidence of abnormal and small pigs in litters
- anaemia

Poultry
- Poor feathering and feather fraying
- Long bones shortened and thickened
- Lesions of tongue and mouth
- Reduced egg production and hatchability
- Affects comb colour

Dogs
- More likely in older animals
- Inherited deficiency seen in Siberian Huskies, Alaskan Malamutes, Doberman Pinschers and Great Danes etc

Horses
- Skin lesions,
- Hair loss with skin scaling on lower limbs and progresses upwards

Zinc

27d

| Zn | Atomic Wt 65.37 | Atomic No 30 | Micro Mineral |

Toxicity

General
- Excess zinc in the diet will lead to feed refusal rather than toxic effects
- Reduced body weight
- Young animals more susceptible than older animals

Ruminants
- Moderate levels are not toxic to ruminants. However at 900 mg/kg of diet may affect absorption and metabolism of copper
- 500ppm in milk replacer for 30 days is toxic to pre-ruminant calves. High Zn interferes with Ca metabolism

Pigs
- Dependent on form of Zn. Lactate and carbonate more toxic than oxide form

Poultry
- Anaemia
- Reduced egg production
- Affected by composition of diet

Dogs
- Anorexia, intermittent vomiting, weakness, depression, hemolytic anaemia with watery diarrhoea
- Ingestion of Zinc ointments can lead to toxic levels

Horses
- Foals more susceptible than adults.
- Swelling, stiffness, lameness, anaemia, unthrifty in young growing foals

Maximum Dietary Tolerable Level (ppm DM)

Cattle	500	Rabbits	500
Sheep	300	Dogs	1000
Pigs	1000	Horses	500
Poultry	1000		

Interrelationships
- Cu, Cd, Zn, phytate, Ca, Fe, fibre, P and Cr

Antagonists
- Absorption adversely affected by calcium (non-ruminants) oxalates, copper and high fibre diets
- Phytate binds zinc in plants and reduces the absorption of zinc in calves and non-ruminant animals but does not affect ruminants
- Calves on milk containing 50% soya protein can reduce zinc absorption to 25% due to phytic acid
- Copper levels would need to be 50 times zinc to cause interference of absorption
- Cadmium is a zinc antimetabolite
- Lead stops absorption of zinc and interferes with the function of zinc
- Iron use can be affected by excess zinc leading to induce anaemia
- Interferes with the utilisation of copper, iron and other trace elements

Synergy
- Zn is necessary for Vitamin A metabolism and transportation
- Organic chelators of zinc can increase the efficiency of absorption of zinc
- Some dietary ingredients, e.g. casein, distillers dried solubles, corn oil can increase Zn absorption
- Cobalt and cadmium have been linked to zinc, also magnesium and nickel
- Additional dietary zinc may help alleviate lead and cadmium toxicity

CONTEXT   www.contextbookshop.com

Zinc

Zn | Atomic Wt 65.37 | Atomic No 30 | Micro Mineral

Main Supplements

Source	Element %	Relative Bio-avail.	Comments
Zinc carbonate	52	med	
Zinc chloride	48	high	
Zinc methionine	4-10	high	
Zinc oxide	46-73	high	insoluble in water low hygroscopicity Pb,As,Cd must be removed. Al or Cl must be <0.5
Zinc proteinate	9-14	high	
Zinc sulphate	22-36	high	high water solubility high hygroscopicity Cd<450ppm
Zinc lysine		high	

CONTEXT — www.contextbookshop.com

Zinc

27f

| Zn | Atomic Wt 65.37 | Atomic No 30 | Micro Mineral |

Feed Name	mg/kg DM
Alfalfa meal	17.7
Bakery waste	16.0
Barley grain	25.6
Bean field	51.0
Blood meal	5.0
Brewers grains	72.2
Buckwheat grain	10.0
Buttermilk dehyd.(cattle)	44.0
Casein dehyd. (cattle)	30.0
Cassava tubers dehy	17.0
Citrus pulp dried	15.0
Copra meal	61.0
Cottonseed Whole	35.2
Cottonseed meal	50.0
Distillers grains - wheat	48.3
Distillers grains maize	55.6
Distillers grains- barley	60.4
Fishmeal (Sth Am)	92.9
Grass bluegrass	-
Grass alfalfa	18.0
Grass bermuda	-
Grass clover	-
Grass extensive	27.0
Grass kikuyu	31.0
Grass timothy	-
Groundnut ext	50.0
Hay alfalfa	24.4
Hay bluegrass	-
Hay clover	17.0
Hay eragostus	7.0
Hominy feed	33.3
Linseed meal (mech ext)	66.0
Maize bran	-
Maize germ ext(sol)	97.0
Maize gluten 20	13.1
Maize gluten 60	72.2
Maize grain	40.0
Malt culms	85.0
Milk (cattle-dehyd)	85.0
Milk skimmed	42.0
Millet grain	23.0
Molasses -beet	20.0

Feed Name	mg/kg DM
Molasses -cane	23.0
Oat groats	-
Oat middlings	152.0
Oatfeed	22.0
Oats grain	31.0
Palm kernel exp	49.5
Peas	35.0
Potato dried	16.6
Rape ext (mech)	61.1
Rice bran	42.0
Rice grain	17.0
Rye grain	36.0
Safflower ext. solv.	44.0
Sesame ext mech	108.0
Silage alfalfa	-
Silage grass	30.0
Silage maize	15.0
Silage sorghum	32.0
Silage wholecrop	21.0
Sorghum grain	19.0
Soya ext.solv	56.2
Soya flour	-
Soya hipro	62.5
Straw barley	12.5
Straw oat	6.5
Straw wheat	6.8
Sugar beet pulp (dehyd)	1.0
Sugar beet pulp (mol)	2.0
Sunflower ext	113.0
Triticale grain	39.0
Wheat (caustic)	29.1
Wheat bran	97.5
Wheat feed	67.4
Wheat germ ext.	145.0
Wheat grain	29.1
Whey low lactose	8.0
Whey(cattle dehyd)	3.0
Yeast (brewers dehyd)	41.0
Yeast (torula dehyd)	100

CONTEXT www.contextbookshop.com

Bioplex® Zinc from Alltech

Zn | Atomic Wt 65.38 | Atomic No 30 | Micro Mineral

Introduction
- Bioplex® Zinc is an organic trace mineral proteinate for use in livestock feeds.

Concentrations Available

Product	Guaranteed Analysis	Ingredients
Bioplex® Zn 10%*	Minimum 10% Zinc	Zinc proteinate
Bioplex® Zn 15%*	Minimum 15% Zinc	Zinc proteinate

*Not all mineral concentrations are available in every country. Contact your local Alltech representative for details.

Physical Characteristics

Appearance
Bioplex® Zinc is a beige powder with no discernible odour.

Storage
Bioplex® Zinc should be stored in a closed container in a cool, dark area. Shelf life under these conditions is 36 months.

Packaging
Bioplex® Zinc is available in 25 kg bags.

Inclusion Rates of Bioplex® Minerals
- Every species requires a different inclusion rate. Also inclusion depends on the motivation for organic supplementation. There are two issues to be examined.
- Organic supplementation can be for performance reasons, here research shows that supplementation with Bioplex® minerals at rates between 20-50% of inorganic minerals provides superior performance effects.
- Supplementation may be driven by the requirement to meet environmental regulations. Bioplex® Zinc can be used as a complete replacement of inorganic salts at a significantly lower inclusion level.

For inclusion recommendations for your region and other details, contact your local Alltech representative.

Benefits (Conditions responsive to improved Zinc status)
- Zinc plays a key role in the correct functioning of enzymes in most major metabolic pathways
- Key role in hormonal system
- Hoof health and skin, hair and feather integrity
- Affects the development of cartilage and bone
- Production and regeneration of keratin - having a direct effect on the integrity of the udder lining and the protection of the mammary gland

Some Symptoms of Zinc Deficiency
- Dermal/skin disorders
- Emaciation
- Impaired sexual function
- Retarded growth
- Delayed wound healing

CONTEXT Directory Order Form (www.contextbookshop.com)

Please send me ____ copy/copies of The MINERALS Directory at the offer price of £30.00 each. (Normal price £35.00)
Please send me ____ copy/copies of The FEEDS Directory Commodity Products Guide at the offer price of £30.00 each. (Normal price £35.00)
Please send me ____ copy/copies of The FORAGES Directory at the offer price of £30.00 each. (Normal price £35.00)
Please send me ____ copy/copies of The VITAMINS Directory at only £35.00 each. (Out April 2007)

Mr/Mrs/Miss/Ms/Dr ____ Initials ____ Surname ____

Job title ____ Company ____

Address ____

____ Postcode ____

Daytime Tel ____ Fax ____ email address ____

☐ I enclose a cheque for £ ____ made payable to Context

☐ Please bill my VISA/Mastercard/Switch Card by £ ____

Card No. ☐☐☐☐ ☐☐☐☐ ☐☐☐☐ ☐☐☐☐   3 Digit Security Code ☐☐☐

Expiry Date ☐☐ Expiry Date ☐☐

Signature ____ Date ____

Please Tick
☐ Farmer
☐ Adviser
☐ College
☐ Trade
☐ Other ____

Post and packaging per book:
UK £ 2.50 ____
Europe £ 4.00 ____
Rest of World £ 6.00 ____
Total ____

CONTEXT

Return to: CONTEXT Products Ltd, 53 Mill Street, Packington, Ashby-de-la-Zouch, Leics. LE65 1WN. Tel: +44 (0) 1530 411337. Fax: +44(0)1530 411289.